SpringerBriefs in Mathematics

SpringerBriefs in Mathematics showcases expositions in all areas of mathematics and applied mathematics. Manuscripts presenting new results or a single new result in a classical field, new field, or an emerging topic, applications, or bridges between new results and already published works, are encouraged. The series is intended for mathematicians and applied mathematicians.

More information about this series at http://www.springer.com/series/10030

SBMAC SpringerBriefs

The **SBMAC SpringerBriefs** series publishes relevant contributions in the fields of applied and computational mathematics, mathematics, scientific computing, and related areas. Featuring compact volumes of 50 to 125 pages, the series covers a range of content from professional to academic.

The Sociedade Brasileira de Matemática Aplicada e Computacional (Brazilian Society of Computational and Applied Mathematics, SBMAC) is a professional association focused on computational and industrial applied mathematics. The society is active in furthering the development of mathematics and its applications in scientific, technological, and industrial fields. The SBMAC has helped to develop the applications of mathematics in science, technology, and industry, to encourage the development and implementation of effective methods and mathematical techniques for the benefit of science and technology, and to promote the exchange of ideas and information between the diverse areas of application.

http://www.sbmac.org.br/

Sebastià Xambó-Descamps

Real Spinorial Groups

A Short Mathematical Introduction

Springer

Sebastià Xambó-Descamps
Universitat Politècnica de Catalunya
Barcelona, Spain

ISSN 2191-8198 ISSN 2191-8201 (electronic)
SpringerBriefs in Mathematics
ISBN 978-3-030-00403-3 ISBN 978-3-030-00404-0 (eBook)
https://doi.org/10.1007/978-3-030-00404-0

Library of Congress Control Number: 2018958521

This Springer imprint is published by the registered company Springer Nature Switzerland AG
The registered company address is: Gewerbestrasse 11, 6330 Cham, Switzerland

Preface

This brief is a write-up of materials presented at lectures and short introductory courses delivered by the author in the last few years. Its purpose is to offer an axiomatic presentation of geometric algebra and to indicate how it is applied to the study of several groups associated to a real (regular) orthogonal space. Dubbed as "spinorial" in the title, they include the versor, pinor, spinor, and rotor groups.

Since the most basic mathematical concepts we need are summarized in the first chapter, our treatment is not only elementary but essentially self-contained and rigorous, and hence it should be useful to people having mastered basic undergraduate mathematics courses, including some familiarity with vector spaces and linear maps (linear algebra).

Our approach provides a quick access to the key concepts and structures, sheds light on the interdependencies of statements, and provides resources for composing proofs, and clues about how to increase their transparency. It may thus be a handy companion reference for the study of a rich literature formed by a wealth of theoretical or applied works on a broad range of topics (see Sects. 6.3–6.5), and thus it could be adequate material for courses offered to advanced undergraduate or early graduate students in mathematics, physics, or engineering.

Here is a quick description of the contents beyond the first chapter. Chapter 2 is devoted to a study of Grassmann's algebra, which in this brief is considered to consist of the exterior algebra endowed with the inner product and the parity and reverse involutions.

The axiomatic presentation of geometric algebra is spelled out in Chap. 3. Here the stress rests, after stating the axioms and their significance, in the introduction and study of a constellation of concepts and results that emerge from those axioms. They emphasize the canonical linear grading, the outer and inner products, the parity and reversion involutions, and a collection of formulas that relate these elements and facilitate their effective manipulation.

The aim of Chap. 4 is to explore how the geometric algebra of an orthogonal space encodes its isometry group. After some preliminaries that include a complete proof of the Cartan-Dieudonné theorem, the (Lipschitz) group of versors of a geometric algebra is introduced, as well as the subgroups of pinors, spinors, and rotors. The fundamental relations that they enjoy with the orthogonal group (versors and pinors), and with the special orthogonal group (spinors and rotors), are established. The chapter ends with an illustration of the formalism in the case of the Euclidean space.

Chapter 5 is intended to deepen the understanding of the groups introduced in Chap. 4, particularly the rotor groups. After establishing, in dimensions up to 5, a simple criterion for an even multivector to be a rotor, several examples are worked out in detail in Sect. 5.1. These include the rotors in dimensions 1 and 2 (which have algebraic and topological properties that are somewhat special with respect to the general case) and the rotors for spaces of dimension 3, which already enjoy properties akin to those that are valid in higher dimensions. Then we turn, in Sect. 5.2, to the study of plane rotors (they are the product of two linearly independent unit vectors of the same signature); establish their classification into elliptic, parabolic, and hyperbolic rotors; and express them as exponentials of bivectors. The section ends with a proof that any rotor in dimension n is the product of at most $n/2$ plane rotors, which can be regarded as a Cartan-Dieudonné theorem for rotors. The last two sections are devoted to the study of infinitesimal rotations, Sect. 5.3; infinitesimal rotors, Sect. 5.4; and the relation between both concepts. The main results are the identifications of infinitesimal rotations as skew-symmetric endomorphisms, of infinitesimal rotors as bivectors, and the natural, albeit quite subtle, isomorphism between the two. These ideas lead naturally to the question of whether any isometry is the exponential of a skew-symmetric endomorphism. We end the chapter with a detailed afirmative proof for the nth dimensional Euclidean space.

Chapter 6, the last, has one section for each of the preceding chapters. It provides supplementary notes, ideas, results, discussions, and bibliographical references. Together with the background material of Chap. 1, and the Exercises section at the end of each chapter, its role is to facilitate the study of the whole, or of meaningful parts, at a pace set according to the reader's experience.

1 ⟩ **2→3→4→5** *Exercises* **6** ⟩

An important remark in this sense is that the exercises are for the most part not unlike other statements in the text, except that more details are left for the reader. They may cover materials that round some aspect of the main text, occasionally including suggestions for streamlined proofs of some standard results, or to highlight interesting applications. Often hints are appended to suggest a possible path to a complete solution. Since the meaning and significance of most exercises is explained in the text, in general it may be advisable to work on them after reading the corresponding chapter.

Conventions Formulas are numbered sequentially in each chapter. Thus Eq. (2.3) denotes the third equation in Chap. 2. Other reusable blocks are numbered sequentially for each section, so that 2.1.3 refers to the third numbered block in Sect. 2.1. Similar references to materials in Chap. 6 are distinguished in a form like ▹ 6.0.1 to signal their supplemental character. The exercises at the end of each chapter are identified in the form E.2.5, which means the fifth exercise in Chap. 2. In the text, these exercises are referred by their identifier and in general the page on which they can be found.

We use expressions of the form $L := R$ to denote that the left-hand side L *is defined as* the value of expression R on the right-hand side.

Another point is that we allow ourselves to use the functional programming convention of writing $f\boldsymbol{x}$ instead of $f(\boldsymbol{x})$ when there is no doubt from the context that f stands for a map and $\boldsymbol{x}$ for its argument. We also consider sequences as a data structure, and so we will use expressions such as $s = s_1, \ldots, s_k$, which for most authors would be specified as $s = (s_1, \ldots, s_k)$. The concatenation of two sequences s and s' is denoted by s, s' (the *comma operator*).

Comments and Errata Comments about corrections or suggestions for improving this brief are very welcome. Please send them to sebastia.xambo@upc.edu, with the subject "Real spinorial groups." With anticipated thanks, they will be collected and acknowledged, together with a list of currently known errata, at the web page https://mat-web.upc.edu/people/sebastia.xambo/GA/brief-spins.html

Acknowledgments The content of this brief is a distillate of some materials presented, between March of 2015 and August of 2016, in two short courses given in the University of San Luis Potosí (SLP, Mexico), in two short courses taught at the Institute of Mathematics of the University of Valladolid (IMUVA, Spain), and in the lectures taught at the 17th "Lluís Santaló" Research Summer School, [149] (▷ 6.0, p. 107 for more details about these and other similar activities). It has also profited from the writing of the first three chapters of [98]. Our thanks to all the people and institutions that made all those activities possible: José Antonio Vallejo (SLP), Félix Delgado de la Mata (director of the IMUVA), Antonio Campillo (the president of the Real Sociedad Matemática Española (RSME) that entrusted me the organization of the 2016 Santaló Research Summer School, which was held at the University Menéndez Pelayo in Santander), and the coauthors of [98], Carlile Lavor (Campinas, São Paulo, Brazil) and Isiah Zaplana (Institute of Organization and Control, Universitat Politècnica de Catalunya, Barcelona, Spain), who in addition provided valuable feedback on drafts of this work. Thanks also to LA GACETA of the RSME for allowing to take the paper [147] as a seed for this text. And thanks to the most deserving person, Elionor Sedó Torres, for all her sustained love and caring support.

Barcelona, Spain
May 2018

Sebastià Xambó-Descamps

Contents

Chapter 1
Mathematical Background

In this chapter we collect mathematical notions and results that are needed later on. After reviewing some general basic notions and facts about groups in Sect. 1.1, and on linear and multilinear algebra in Sect. 1.2, we study metrics on a real vector space, Sect. 1.3, and finally, in Sect. 1.4, the last, we recall what we need about algebras.

Notations and Conventions

A map f from a set X to a set Y is often specified by sentences such as "the map $f : X \to Y$ such that $x \mapsto f(x)$," where in each case $f(x) \in Y$ is an expression or description of the image of $x \in X$ by f.

The set of ordered pairs (x, y) of elements $x \in X$ and $y \in Y$ is denoted $X \times Y$ and is called the *Cartesian product* of X and Y.

A map $X \times X \to X$, $(x, x') \mapsto x * x'$, is called a *binary operation* on X.

1.1 General Notions on Groups

A *group* is a non-empty set G endowed with a binary operation

$$G \times G \to G, \quad (a, b) \mapsto a * b,$$

which is *associative* (so $(a * b) * c = a * (b * c)$ for all $a, b, c \in G$), has a *neutral element* e (so $a * e = e * a = a$ for all $a \in G$), and for each element $a \in G$ there exists $a' \in G$ such that $a * a' = a' * a = e$. The neutral element is unique and for each $a \in G$, the element a' is also unique. The element a' is called the *inverse* of a and is usually denoted a^{-1}. Note that $(ab)^{-1} = b^{-1}a^{-1}$, as

$$(b^{-1}a^{-1})(ab) = b^{-1}(a^{-1}a)b = e.$$

S. Xambó-Descamps, *Real Spinorial Groups*, SpringerBriefs in Mathematics,
https://doi.org/10.1007/978-3-030-00404-0_1

1.1.1 (Example: The Permutation Group $\Pi(X)$) If X is a set, the set $\Pi(X)$ of all bijective maps $X \to X$ (which are also called *permutations* of X) is a group with operation the *composition of maps* (denoted by juxtaposition of its operands), namely

$$(ab)(x) = a(b(x)) \text{ for all } x \in X \text{ and } a, b \in \Pi(X).$$

Its neutral element is the *identity map* $\mathrm{Id}_X : X \to X$, and the inverse of $a \in \Pi(X)$ is the inverse a' of a as a bijection, which means that $a'(x)$ is, for each $x \in X$, the unique element of X such that $a(a'(x)) = x$.

For any positive integer n, the permutation group of the set $N = \{1, \ldots, n\}$ is denoted by S_n. Its cardinal is $n!$. If j is a permutation, instead of $j(k)$ we usually write j_k and j itself is often denoted by $[j_1, \ldots, j_n]$.

The permutation that consists in swapping $j, k \in N$ $(j < k)$ is denoted (j, k) and it is called the *transposition* of j and k. As it is easily argued, any permutation can be expressed as the composition of a finite number of transpositions of the form $(j, j + 1)$ $(1 \leqslant j < n)$. For example, $[3, 2, 1] = (1, 2)(2, 3)(1, 2)$, or also $(2, 3)(1, 2)(2, 3)$.

A permutation is *even* (*odd*) if it can be expressed as the composition of an even (odd) number of transpositions. The *sign* of a permutation $j = [j_1, \ldots, j_n]$ is $(-1)^{t(j)}$, where $t(j)$ is the number of order inversions in j, that is, the number of pairs (r, s) such that $1 \leqslant r < s \leqslant n$ but $j_r > j_s$. For example, $t([3, 2, 1]) = 3$. As it turns out, a permutation is even or odd according to whether its sign $+1$ or -1. In other words, it is even (odd) when $t(j)$ is even (odd).

1.1.2 (Example: Product of Groups) Given two groups G and G', the *product group* of G and G' is defined as the Cartesian product $G \times G'$ endowed with the (component-wise) operation $((g, g'), (h, h')) \mapsto (gh, g'h')$. With this operation, $G \times G'$ is a group: associativity is a direct consequence of the associativity of the group operations of G and G'; if $e \in G$ and $e' \in G'$ are the neutral elements of G and G', then (e, e') is the neutral element of $G \times G'$; and given $g \in G$ and $g' \in G'$, (g^{-1}, g'^{-1}) is inverse of (g, g'). The product $G_1 \times \cdots \times G_k$ of groups $G_1, \ldots, G_k$ is defined in a similar way.

1.1.3 (Subgroups) A non-empty subset H of a group G is a *subgroup* if for all $a, b \in H$ we have $ab \in H$ and $a^{-1} \in H$ (we again use no symbol to denote the group operation). Then it follows directly from the definitions that $e \in H$ and that H is a group with the restriction of the binary operation to H.

Many interesting groups are defined as subgroups of the permutation group $\Pi(X)$ of some X, which in turn are usually defined by imposing some extra properties to the permutations.

For example, the set A_n of even permutations of S_n is a subgroup (it is called *alternating group*).

1.1.4 (Abelian Groups) A group G is called *abelian* (or *commutative*) if $ab = ba$ for all $a, b \in G$. For example, the set of non-zero real numbers is a commutative group with the product of real numbers. We say that it is the *multiplicative* group of non-zero real numbers because we use for it the *multiplicative notation* (the identity element is 1 and the inverse of a positive real number x is $1/x$, which is also called the *reciprocal* of x).

The operation of abelian groups is sometimes written in *additive notation* $a+b$ (in which case the commutativity is expressed as $a+b=b+a$), the identity element is denoted by 0 (thus $a+0=0+a=a$ for all $a\in G$), and the inverse a' is denoted $-a$ (it is called the *opposite* of a and so, by definition, $a+(-a)=(-a)+a=0$ for all $a\in G$). For example, the real numbers with the sum is a commutative *additive* group, which just means that the additive notation is being used.

1.1.5 (Homomorphisms) A map $f:G\to G'$ between the groups G and G' is said to be a *group homomorphism* if $f(xy)=f(x)f(y)$ for all $x,y\in G$. This definition implies that $f(e)=e'$, where e' is the neutral element of G', and that $f(x^{-1})=(f(x))^{-1}$ for all $x\in G$.

The *kernel* of f is defined as

$$\ker(f)=\{x\in G: f(x)=e'\}.$$

It is a subgroup of G. This subgroup satisfies that $yxy^{-1}\in\ker(f)$ for all $x\in\ker(f)$ and all $y\in G$, a fact that is expressed by saying that it is a *normal* (or *invariant*) subgroup.

1.1.6 (Quotient Group) A *quotient* of a group G by a normal subgroup H is an onto homomorphism $\pi:G\to\bar{G}$ such that $\ker(\pi)=H$.

An example is the sign map $\sigma:S_n\to\{\pm1\}$, which is a quotient of S_n by the alternating subgroup A_n. Indeed, it is immediate that it is an onto homomorphism and that its kernel is the alternating subgroup A_n.

The main property of a quotient $\pi:G\to\bar{G}$ by H is the following *universal property*:

If $f:G\to G'$ is a homomorphism such that $H\subseteq\ker(f)$, then there exists a unique group homomorphism $\bar{f}:\bar{G}\to G'$ such that $\bar{f}(\pi x)=f(x)$ for all $x\in G$.

Proof First note that if $\pi(x)=\pi(x')$, then $z=x^{-1}x'\in\ker(\pi)=H$, so $x'=xz$ and

$$f(x')=f(xz)=f(x)f(z)=f(x).$$

This says that the map $\bar{f}:\pi(x)\mapsto f(x)$ is well-defined, for any element $\bar{x}\in\bar{G}$ can be written in the form $\bar{x}=\pi(x)$, $x\in G$, and the value of $f(x)$ does not depend on which x we choose. This shows that $\bar{f}$ is uniquely determined and it is straightforward to check that it is a group homomorphism:

$$\bar{f}(\pi(x)\pi(x'))=\bar{f}(\pi(xx'))=f(xx')=f(x)f(x')=\bar{f}(\pi x)\bar{f}(\pi x').$$

□

If $\pi : G \to \bar{G}$ and $\pi' : G \to \bar{G}'$ are two quotients of G by the normal subgroup H, then there is a unique group homomorphism $\bar{\pi}' : \bar{G} \to \bar{G}'$ such that $\bar{\pi}'(\pi x) = \pi'(x)$, and a unique group homomorphism $\bar{\pi} : \bar{G}' \to \bar{G}$ such that $\bar{\pi}(\pi' x) = \pi(x)$, for all $x \in G$ (the preceding statement applies to both π' and π). It follows that $\bar{\pi}'\bar{\pi}$ and $\bar{\pi}\bar{\pi}'$ are the identity of $\bar{G}$ and $\bar{G}'$, respectively, and hence $\bar{\pi}$ and $\bar{\pi}'$ are inverse isomorphisms. This proves that if there is a quotient, then it is unique up to a natural isomorphism.

The construction of a quotient is easily obtained with the ideas of the proof above. We note that there is bijection of the set G/H of subsets of the form $xH = \pi^{-1}(\pi x)$, $x \in G$, and $\bar{G}$, namely $xH \mapsto \pi(x)$. This suggests to define the product in G/H (its elements are called the *left cosets* of H) by the rule $(xH)(x'H) = (xx')H$, and then it is immediate to check that G/H is a group and that the map $\pi : G \to G/H$, $x \mapsto xH$, is a quotient of G by H. We will say that it is *the* quotient of G by H, and by the above we know that we have a natural isomorphism $G/H \simeq \bar{G}$, $xH \mapsto \pi(x)$, for any quotient $\pi : G \to \bar{G}$.

So we may write $S_n/A_n \simeq \{\pm 1\}$. Or, more generally,

$$G/\ker(f) \simeq f(G) \tag{1.1}$$

for any group homomorphism $f : G \to G'$.

1.1.7 (Finite Groups) A group G is finite, if it has finitely many elements, and in this case its cardinal $n = |G|$ is called the *order* of G.

It is not difficult to see that if G is finite, and H is a normal subgroup of G, then $|G| = |H| \times |G/H|$ (see E.1.1, p. 18).

The set $\mathbb{Z}_n = \{0, 1, \ldots, n-1\}$ of possible remainders of the integer division by a positive integer n is an abelian group of order n with the sum $a +_n b$ defined as the remainder of the division by n of the ordinary sum $a + b$, a result that is also expressed as $a + b \bmod n$. In $\mathbb{Z}_2$, for example, we have $1 +_2 1 = 0$, or $1 + 1 \bmod 2 = 0$, or also $1 + 1 \equiv 0 \bmod 2$. It is easy to see that any group of order 2 is isomorphic to $\mathbb{Z}_2$, and that any group of order 3 is isomorphic to $\mathbb{Z}_3$, but there are, up to isomorphism, two groups of order 4, namely $\mathbb{Z}_4$, which occurs if there is an element x such that $x^2 \neq e$, and $\mathbb{Z}_2 \times \mathbb{Z}_2$, which occurs when $x^2 = e$ for any of its four elements x. The latter group is usually called the *Klein group* of order 4.

1.2 Linear and Multilinear Algebra

By a *vector space* we understand an $\mathbb{R}$-vector space unless indicated otherwise. Recall that this structure consists of an abelian (additive) group E, whose elements are called *vectors*, and which we will denote by bold italic symbols $(\boldsymbol{e}, \boldsymbol{u}, \boldsymbol{v}, \boldsymbol{x}, \boldsymbol{y}, \ldots)$, together with a map

$$\mathbb{R} \times E \to E, \quad (\lambda, \boldsymbol{x}) \mapsto \lambda \boldsymbol{x}$$

such that

$$\lambda(\boldsymbol{x}+\boldsymbol{x}') = \lambda\boldsymbol{x} + \lambda\boldsymbol{x}'$$
$$(\lambda+\lambda')\boldsymbol{x} = \lambda\boldsymbol{x} + \lambda'\boldsymbol{x}$$
$$\lambda(\lambda'\boldsymbol{x}) = (\lambda\lambda')\boldsymbol{x}$$
$$1\boldsymbol{x} = \boldsymbol{x}$$

for all $\boldsymbol{x}, \boldsymbol{x}' \in E$, $\lambda, \lambda' \in \mathbb{R}$.

A *linear combination* of the vectors $\mathbf{e} = \boldsymbol{e}_1, \ldots, \boldsymbol{e}_n$ is an expression of the form

$$\lambda_1\boldsymbol{e}_1 + \cdots + \lambda_n\boldsymbol{e}_n,$$

where $\lambda_1, \ldots, \lambda_n \in \mathbb{R}$.

1.2.1 (Bases and Dimension) The set of linear combinations of $\mathbf{e}$ is denoted $\langle \boldsymbol{e}_1, \ldots, \boldsymbol{e}_n \rangle$. It is a *vector subspace* of E, which means that it is closed with respect to the sum of vectors and with respect to the product of a scalar times a vector.

If $\langle \boldsymbol{e}_1, \ldots, \boldsymbol{e}_n \rangle = E$, we say that the vectors $\mathbf{e}$ *span* E. If the relation

$$\lambda_1\boldsymbol{e}_1 + \cdots + \lambda_n\boldsymbol{e}_n = 0$$

only happens for $\lambda_1 = \cdots = \lambda_n = 0$, we say that the vectors $\mathbf{e}$ are *linearly independent* (or *linearly dependent* otherwise). And $\mathbf{e}$ is a (linear) *basis* of E if the vectors are linearly independent and span E. In this case the integer n has the same value for all bases, is called the *dimension* of E, and is denoted by $\dim(E)$. Given a vector $\boldsymbol{x}$, the scalars $\lambda_1, \ldots, \lambda_n$ such that $\boldsymbol{x} = \lambda_1\boldsymbol{e}_1 + \cdots + \lambda_n\boldsymbol{e}_n$ are called the *components* of $\boldsymbol{x}$ with respect to the basis $\mathbf{e}$.

1.2.2 (Example: The Supermarket Vector Space) Let J be a finite set, which we may think as the set of goods on sell by a supermarket. Consider the set $\mathbb{R}[J]$ formed by all the maps $x : J \to \mathbb{R}$ (we may think of any such x as the concrete shopping of an amount $x(j)$ of the j-th good). For $x, x' \in \mathbb{R}[J]$ and $\lambda \in \mathbb{R}$, define $x + x'$ and λx by the rules

$$(x+x')(j) = x(j) + x'(j), \quad (\lambda x)(j) = \lambda x(j).$$

With this operation, it is straightforward to check that $\mathbb{R}[J]$ is a vector space.

If for any given $j \in J$ we define $\delta_j \in \mathbb{R}[J]$ as the map such that $\delta_j(j) = 1$ and $\delta_j(j') = 0$ for any $j' \neq j$ (so δ_j is the shopping of one unit of j and nothing else), then we have the identity $x = \sum_{j\in J} x(j)\delta_j$ and this implies that $\{\delta_j\}_{j\in J}$ is a basis of $\mathbb{R}[J]$. In particular, $\dim \mathbb{R}[J] = |J|$.

We can define a similar vector space for any set J (not necessarily finite), but in this case the maps $x : J \to \mathbb{R}$ are restricted by the condition that $x(j) = 0$

for all j but a finite number. This condition guarantees that the decomposition $x = \sum_{j\in J} x(j)\delta_j$ is still valid (the sum only involves a finite number of non-zero terms).

1.2.3 (Example: E^k) For any positive integer k, the set E^k of k-tuples

$$\mathbf{x} = (\boldsymbol{x}_1, \ldots, \boldsymbol{x}_k), \quad \boldsymbol{x}_1, \ldots, \boldsymbol{x}_k \in E,$$

is a vector space with the sum $\mathbf{x} + \mathbf{y}$ defined, if $\mathbf{y} = (\boldsymbol{y}_1, \ldots, \boldsymbol{y}_k) \in E^k$, as the k-tuple

$$(\boldsymbol{x}_1 + \boldsymbol{y}_1, \ldots, \boldsymbol{x}_k + \boldsymbol{y}_k),$$

and the product $\lambda\mathbf{x}$, $\lambda \in \mathbb{R}$, as

$$(\lambda\boldsymbol{x}_1, \ldots, \lambda\boldsymbol{x}_k).$$

We have $\dim E^k = kn$ ($n = \dim E$), because if $\boldsymbol{e}_1, \ldots, \boldsymbol{e}_n$ is a basis of E, then the k-tuples $\mathbf{e}_{ij} = (0, \ldots, 0, \boldsymbol{e}_j, 0, \ldots, 0)$, $1 \leqslant j \leqslant n$, and $\boldsymbol{e}_j$ in any position $i = 1, \ldots, k$, form a basis of E^k.

Linear Maps

A map $f : E \to E'$ of the vector space E to the vector space E' is said to be *linear* if

$$f(\boldsymbol{x} + \boldsymbol{x}') = f(\boldsymbol{x}) + f(\boldsymbol{x}') \text{ and } f(\lambda\boldsymbol{x}) = \lambda f(\boldsymbol{x})$$

for all $\boldsymbol{x}, \boldsymbol{x}' \in E$ and $\lambda \in \mathbb{R}$.

Linear maps are often specified by the following principle:

1.2.4 (Construction of Linear Maps) *Given any vectors $\boldsymbol{e}'_1, \ldots, \boldsymbol{e}'_n \in E'$, there exists a unique linear map $f : E \to E'$ such that $f(\boldsymbol{e}_j) = \boldsymbol{e}'_j$ ($j = 1, \ldots, n$).* □

Note that the map in the preceding statement is given by the formula

$$f(\lambda_1\boldsymbol{e}_1 + \cdots + \lambda_n\boldsymbol{e}_n) = \lambda_1\boldsymbol{e}'_1 + \cdots + \lambda_n\boldsymbol{e}'_n.$$

The *kernel* of a linear map f,

$$\ker(f) = \{\boldsymbol{x} \in E : f(\boldsymbol{x}) = 0\},$$

is a vector subspace of E. For a linear map to be injective, it is clearly necessary that $\ker(f) = \{0\}$, and it is easy to see that this condition is also sufficient. The *image* of f is the linear subspace $f(E)$ of E' defined by

$$f(E) = \{f(\boldsymbol{x}) : \boldsymbol{x} \in E\}.$$

Here is the fundamental relation between the dimensions of $f(E)$ and $\ker(f)$:

$$\dim(f(E)) = \dim(E) - \dim(\ker(f)). \quad (1.2)$$

A linear map $f : E \to E'$ is a *linear isomorphism* if it is one-to-one and onto, or, in other words, if $\ker(f) = \{0\}$ and $f(E) = E'$.

The set $\mathrm{Lin}(E, E')$ of linear maps from E to E' is itself a vector space with the operations $f + g$ and λf defined by

$$(f + g)(\boldsymbol{x}) = f(\boldsymbol{x}) + g(\boldsymbol{x}) \text{ and } (\lambda f)(\boldsymbol{x}) = \lambda f(\boldsymbol{x}).$$

The dimension of this space is (if $n' = \dim(E')$)

$$\dim \mathrm{Lin}(E, E') = nn' = \dim(E)\dim(E'). \quad (1.3)$$

This is an easy consequence of 1.2.4, as it shows that there is a linear isomorphism $\mathrm{Lin}(E, E') \simeq E'^n$ (this isomorphism depends on choosing a basis $\boldsymbol{e}_1, \ldots, \boldsymbol{e}_n$ of E).

1.2.5 (Matrix of a Linear Map) Let $f : E \to E'$ be a linear map. Let $\mathbf{e} = \boldsymbol{e}_1, \ldots, \boldsymbol{e}_n$ be a basis of E and $\mathbf{e}' = \boldsymbol{e}'_1, \ldots, \boldsymbol{e}'_{n'}$ a basis of E'. For each $j = 1, \ldots, n$, express $f\boldsymbol{e}_j$ as a linear combination of $\mathbf{e}'$, say

$$f\boldsymbol{e}_j = \alpha_j^1 \boldsymbol{e}'_1 + \cdots + \alpha_j^{n'} \boldsymbol{e}_{n'}. \quad (*)$$

The matrix $A = (\alpha_j^k)$ $(1 \leqslant j \leqslant n, 1 \leqslant k \leqslant n')$ is the *matrix of* f *with respect to the bases* $\mathbf{e}$ *and* $\mathbf{e}'$. Notice that the relations in $(*)$ can be written in a compact matrix form as

$$f\mathbf{e} = \mathbf{e}'A,$$

where we regard $\mathbf{e}$ and $\mathbf{e}'$ as one-row matrices (of vectors). Indeed: the expression $\mathbf{e}'A$ is a one-row matrix (of vectors); its j-th entry is $\mathbf{e}'A_j$, where A_j is the j-th column of A; the entries of A_j are precisely the coefficients α_j^k $(k = 1, \ldots, n')$ in the relation $(*)$; therefore $\mathbf{e}'A_j = f\boldsymbol{e}_j$, which is the j-th entry of $f\mathbf{e}$.

Quotient Space

1.2.6 If F is a linear subspace of E, the quotient group E/F has a unique vector space structure for which the quotient map $\pi : E \to E/F$ is linear. In this context, the universal property of the quotient says that for any linear map $f : E \to E'$ such that $F \subseteq \ker(f)$, there exists a unique linear map $\bar{f} : E/F \to E'$ such that $\bar{f}(\pi \boldsymbol{x}) = f(\boldsymbol{x})$ for all $\boldsymbol{x} \in E$. We will say that E/F with this structure is the *quotient* of E by F.

1.2.7 Let F' be a vector subspace of a space E' and $f : E \to E'$ a linear map such that $f(F) \subseteq F'$. Then there exists a unique linear map

$$\bar{f} : E/F \to E'/F'$$

such that $\bar{f}(\pi \boldsymbol{x}) = \pi'(f\boldsymbol{x})$, where π and π' are the corresponding quotient maps. Indeed, the kernel of the composition $E \xrightarrow{f} E' \xrightarrow{\pi'} E'/F'$, $\boldsymbol{x} \mapsto \pi'(f\boldsymbol{x})$, contains F, and so there is a unique linear map $E/F \to E'/F'$ such that $\pi\boldsymbol{x} \mapsto \pi'(f\boldsymbol{x})$.

Dual Space

The space $E^* := L(E, \mathbb{R})$ is called the *dual space* of E. Its elements are called *linear forms* of E, or just 1-*forms*, and will be denoted by boldface Greek characters unless indicated otherwise.

Given a basis $\mathbf{e} = \boldsymbol{e}_1, \ldots, \boldsymbol{e}_n$ of E, a linear form $\boldsymbol{\alpha} : E \to \mathbb{R}$ is specified by giving n scalars $\alpha_1, \ldots, \alpha_n \in \mathbb{R}$ and setting (see 1.2.4)

$$\boldsymbol{\alpha}(\lambda_1\boldsymbol{e}_1 + \cdots + \lambda_n\boldsymbol{e}_n) = \lambda_1\alpha_1 + \cdots + \lambda_n\alpha_n.$$

The *dual basis* of the basis $\boldsymbol{e}_1, \ldots, \boldsymbol{e}_n$ of E is denoted by $\boldsymbol{e}^1, \ldots, \boldsymbol{e}^n$ and is defined by the relations $\boldsymbol{e}^j(\boldsymbol{e}_k) = \delta^j_k$, where δ^j_k is the so-called *Kronecker's symbol*:

$$\delta^j_k = \begin{cases} 0 & \text{if } k \neq j, \\ 1 & \text{if } k = j. \end{cases}$$

With the notations of the previous paragraph, we have $\boldsymbol{\alpha} = \alpha_1\boldsymbol{e}^1 + \cdots + \alpha_n\boldsymbol{e}^n$.

For the space $\mathbb{R}[J]$ studied in Example 1.2.2 (J finite), to give a 1-form $\boldsymbol{\alpha}$ is equivalent to give the numbers $\alpha_j = \boldsymbol{\alpha}(\delta_j)$ ($j \in J$), as

$$\boldsymbol{\alpha}(x) = \boldsymbol{\alpha}\big(\textstyle\sum_{j\in J} x(j)\delta_j\big) = \sum_{j\in J} x(j)\boldsymbol{\alpha}(\delta_j) = \sum_{j\in J} x(j)\alpha_j.$$

In other words, a 1-form $\boldsymbol{\alpha}$ of $\mathbb{R}[J]$ is the same as fixing a price α_j for any of good $j \in J$, and $\boldsymbol{\alpha}(x)$ is the price of x according to that pricing. Here a negative value of α_j maybe thought as the amount payed by the supermarket to buy one unit of j, so that a shopping may accommodate sellings as well as buyings.

1.2.8 (Dual Map) If $f : E \to E'$ is a linear map, its *dual* is the map $f^* : E'^* \to E^*$ defined by the relation $f^*(\boldsymbol{\alpha}')(\boldsymbol{x}) = \boldsymbol{\alpha}'(f\boldsymbol{x})$.

If $\mathbf{e}$ is a basis of E, $\mathbf{e}'$ a basis of E', and A the matrix of f with respect to $\mathbf{e}$ and $\mathbf{e}'$, then the matrix (β^j_k) of f^* (with respect to the dual bases $\mathbf{e}'^*$ and $\mathbf{e}^*$ of $\boldsymbol{e}'$ of $\mathbf{e}$) is A^T (the transpose of A). Notice that $\beta^j_k = (f^*\boldsymbol{e}'^k)(\boldsymbol{e}_j) = \boldsymbol{e}'^k(f\boldsymbol{e}_j) = \alpha^k_j$.

1.2.9 (Example) If F is a linear subspace of E, the dual map $i^* : E^* \to F^*$ is the *restriction map*: $i^*(\boldsymbol{\alpha})$ is the restriction of $\boldsymbol{\alpha}$ to F. This means that for any $\boldsymbol{\alpha} \in E^*$,

$i^*(\boldsymbol{\alpha})$ is the restriction of $\boldsymbol{\alpha}$ to F. Indeed, $i^*(\boldsymbol{\alpha})(\boldsymbol{x}) = \boldsymbol{\alpha}(i\boldsymbol{x}) = \boldsymbol{\alpha}(\boldsymbol{x})$ for all $\boldsymbol{x} \in F$. Since any linear form of F is the restriction of a linear form of E, we get that the restriction map $E^* \to F^*$ is onto (see E.1.2, p. 18, for details).

Multilinear Maps

If k is a positive integer, E^k denotes the vector space of k-tuples $(\boldsymbol{x}_1, \ldots, \boldsymbol{x}_k)$ of vectors of E (see 1.2.3). If E' is another vector space, a map $f : E^k \to E'$ is said to be *multilinear*, or *k-linear* if we want to specify k, if it is linear in each of each arguments when the value of the others is fixed, but arbitrary. For the determination of a multilinear map $E^k \to E'$ we have a principle analogous to 1.2.4:

1.2.4′ *Given vectors* $\boldsymbol{e}'_{j_1,\ldots,j_k} \in E'$ $(1 \leqslant j_1, \ldots, j_k \leqslant n)$ *there is a unique multilinear map* $f : E^k \to E'$ *such that* $f(\boldsymbol{e}_{j_1}, \ldots, \boldsymbol{e}_{j_k}) = \boldsymbol{e}'_{j_1,\ldots,j_k}$. □

The set $\mathrm{Lin}_k(E, E')$ of k-linear maps $E^k \to E'$ is itself a vector space with the operations $f + g$ and λf defined as for the case of linear maps. Note that $\mathrm{Lin}_1(E, F) = \mathrm{Lin}(E, F)$. Using 1.2.4′ we get that

$$\dim \mathrm{Lin}_k(E, E') = n^k n'.$$

A map $f : E^k \to E'$ is said to be *symmetric* if its value does not change when we swap any two consecutive arguments:

$$f(\boldsymbol{x}_1, \ldots, \boldsymbol{x}_{j+1}, \boldsymbol{x}_j, \ldots, \boldsymbol{x}_k) = f(\boldsymbol{x}_1, \ldots, \boldsymbol{x}_j, \boldsymbol{x}_{j+1}, \ldots, \boldsymbol{x}_k) \text{ for } j = 1, \ldots, k-1.$$

Since any permutation can be obtained as a succession of neighboring swaps, a symmetric function remains invariant under any permutation of its arguments. For symmetric k-multilinear maps we have:

1.2.4″ *Given vectors* $\boldsymbol{e}'_{j_1,\ldots,j_k} \in E'$ $(1 \leqslant j_1 \leqslant \ldots \leqslant j_k \leqslant n)$, *there is a unique symmetric multilinear map* $f : E^k \to E'$ *such that* $f(\boldsymbol{e}_{j_1}, \ldots, \boldsymbol{e}_{j_k}) = \boldsymbol{e}'_{j_1,\ldots,j_k}$. □

The set $\mathrm{Sym}_k(E, E')$ of symmetric k-linear maps is a linear subspace of $\mathrm{Lin}_k(E, E')$ and we have

$$\dim \mathrm{Sym}_k(E, E') = \binom{n-k+1}{k-1} n'.$$

A multilinear map $f : E^k \to E'$ is said to be *skew-symmetric* if it changes sign when we swap any two successive arguments:

$$f(\boldsymbol{x}_1, \ldots, \boldsymbol{x}_{j+1}, \boldsymbol{x}_j, \ldots, \boldsymbol{x}_k) = -f(\boldsymbol{x}_1, \ldots, \boldsymbol{x}_j, \boldsymbol{x}_{j+1}, \ldots, \boldsymbol{x}_k) \text{ for } j=1, \ldots, k-1.$$

Under an arbitrary permutation of its arguments, the value of such maps is changed by the sign of the permutation.

For skew-symmetric k-linear maps we have:

1.2.4‴ *Given vectors* $e'_{j_1,\ldots,j_k} \in E'$ $(1 \leqslant j_1 < \ldots < j_k \leqslant n)$ *there is a unique skew-symmetric k-linear map* $f : E^k \to E'$ *such that* $f(e_{j_1}, \ldots, e_{j_k}) = e'_{j_1,\ldots,j_k}$. □

The set $\mathrm{Alt}_k(E, E')$ of skew-symmetric k-linear maps forms a linear subspace of $\mathrm{Lin}_k(E, E')$ and we have

$$\dim \mathrm{Alt}_k(E, E') = \binom{n}{k} n'.$$

The Projective Space P*E*

The projective space $\mathbf{P}E$ of the vector space E is *the set of one-dimensional linear subspaces of* E. To distinguish between the subspace $\langle e\rangle$ ($e \in E-\{0\}$) as a subset of E and as a point of $\mathbf{P}E$, the latter will be denoted by $|e\rangle$ (this is the Dirac *ket* notation and corresponds to the notation $[e]$ often used in projective geometry texts). Thus we have $|e\rangle = |e'\rangle$ if and only if $e' = \lambda e$ for some $\lambda \in \mathbb{R}$ (necessarily non-zero because $e, e' \in E - \{0\}$), a relation that henceforth will be written $e' \sim e$.

1.3 Metrics

With the symbols (E, q) we denote the *quadratic space* (also called an *orthogonal space*) composed of a vector space E and a *metric* q of E, that is, a symmetric bilinear map

$$q : E \times E \to \mathbb{R}.$$

Instead of $q(x, x)$, we will write $q(x)$ and say that it is the *quadratic form* associated with q. Note that $q(\lambda x) = \lambda^2 q(x)$.

A vector x is said to be *q-isotropic*, of *null*, if $q(x) = 0$. Non-null (or non-isotropic) vectors are said to be *positive* or *negative* according to whether $q(x) > 0$ or $q(x) < 0$.

The quadratic form determines the metric through the *polarization formula* (see E.1.3, p. 18):

$$2q(x, y) = q(x + y) - q(x) - q(y). \tag{1.4}$$

The metric q is said to be *regular*, or *non-degenerate*, if $x \in E$ and $q(x, y) = 0$ for all $y \in E$ imply that $x = 0$. Or, in other words, if for each non-zero $x \in E$ there exists $y \in E$ such that $q(x, y) \neq 0$.

Two vectors $x, y \in E$ are said to be *q-orthogonal*, or simply *orthogonal*, when $q(x, y) = 0$. Of a basis $\mathbf{e}$ such that $q(e_j, e_k) = 0$ for all $j \neq k$ we say that it is *orthogonal*. Any metric q admits orthogonal basis (see E.1.4, p. 18).

The *signature* (r, s) of a metric q denotes that in an orthogonal basis $\mathbf{e}$ there are r vectors such that $q(\boldsymbol{e}_j) > 0$ and s vectors such that $q(\boldsymbol{e}_j) < 0$. Note that $r + s \leqslant n$, with equality if and only if q is regular. Indeed, if there exists j such that $q(\boldsymbol{e}_j) = 0$, then we have $q(\boldsymbol{e}_j, \boldsymbol{e}_k) = 0$ for all k and q is degenerate. The definition of (r, s) does not depend on the orthogonal basis used to find r and s (*Sylvester's law*, see E.1.5, p. 19). Any orthogonal basis can be normalized so that $q(\boldsymbol{e}_j) = \pm 1$ for all j such that $q(\boldsymbol{e}_j) \neq 0$. Such bases are called *orthonormal.*

The signature is also denoted by $(r, s, t) = (r, s, n - (r + s))$ if it is wished to make explicit the number t of indexes j such that $q(\boldsymbol{e}_j) = 0$. With this notation, we have the equality $n = r + s + t$.

1.3.1 *If* $\mathbf{e}$ *is orthogonal, and we set* $q_j = q(\boldsymbol{e}_j)$*, then* $q(\boldsymbol{x}, \boldsymbol{y}) = \sum_j q_j \xi_j \eta_j$*, where* $\boldsymbol{x} = \sum_j \xi_j \boldsymbol{e}_j$ *and* $\boldsymbol{y} = \sum_j \eta_j \boldsymbol{e}_j$*. In particular,* $q(\boldsymbol{x}) = \sum_j q_j \xi_j^2$*.*

Proof Since q is bilinear, we have $q(\boldsymbol{x}, \boldsymbol{y}) = \sum_{j,k} \xi_j \eta_k q(\boldsymbol{e}_j, \boldsymbol{e}_k)$. Now it suffices to note that $q(\boldsymbol{e}_j, \boldsymbol{e}_k) = 0$ if $k \neq j$ and $q(\boldsymbol{e}_j, \boldsymbol{e}_j) = q(\boldsymbol{e}_j) = q_j$. □

1.3.2 *If a metric* q' *of a vector space* E' *also has signature* (r, s)*, there exists a linear isomorphism* $f : E \to E'$ *such that* $q'(f\boldsymbol{x}, f\boldsymbol{y}) = q(\boldsymbol{x}, \boldsymbol{y})$ *for all* $\boldsymbol{x}, \boldsymbol{y} \in E$*.*

Proof If $\mathbf{e}'$ is an q'-orthonormal basis of E' and its signature is (r, s), then we may assume, perhaps after reordering $\mathbf{e}'$, that $q_j = q(\boldsymbol{e}_j) = q'(\boldsymbol{e}'_j) = q'_j$ for all j. Now consider the unique linear isomorphism $f : E \to E'$ such that $f(\boldsymbol{e}_j) = \boldsymbol{e}'_j$. Then we have, with the same notations as in 1.3.1 (used also for q'), $f(\boldsymbol{x}) = \sum_j \xi_j \boldsymbol{e}'_j$ and $f(\boldsymbol{y}) = \sum_j \eta_j \boldsymbol{e}'_j$, so $q'(f\boldsymbol{x}, f\boldsymbol{y}) = \sum_j q'_j \xi_j \eta_j$, and this agrees with $q(\boldsymbol{x}, \boldsymbol{y}) = \sum_j q_j \xi_j \eta_j$ because $q'_j = q_j$. □

Let (E, q) and (E', q') be quadratic spaces. An isomorphism $f : E \to E'$ satisfying the relation $q'(f\boldsymbol{x}, f\boldsymbol{y}) = q(\boldsymbol{x}, \boldsymbol{y})$ for all $\boldsymbol{x}, \boldsymbol{y} \in E$ is said to be an *isometry*. With this terminology, we may phrase 1.3.2 by saying that *the signature determines the metric* q *up to an isometry.*

The isometries $f : E \to E$ form a group with the composition operation. This group is called the *orthogonal group* of q, or of (E, q), and will be denoted by $\mathrm{O}_q(E)$. Since it is determined by the signature (r, s) of q up to isomorphism, it will usually be denoted by $\mathrm{O}_{r,s}$. The study of this group by means of geometric algebra is one of the main goals of this brief.

In the next two subsections assume that the metric q is regular.

Orthogonal Space

If F is a subset of E, we define the *orthogonal* $F^\perp$ of F as follows:

$$F^\perp = \{\boldsymbol{x} \in E : q(\boldsymbol{x}, \boldsymbol{y}) = 0 \text{ for all } \boldsymbol{y} \in F\}.$$

The linearity of q implies that $F^\perp$ is a linear subspace of E and that $F^\perp = \langle F \rangle^\perp$.

1.3.3 (1) $\dim(F^{\perp}) = n - \dim(F)$. (2) $F^{\perp\perp} = F$.

Proof

(1) Consider the map $q^{\sharp} : E \to E^*$ given by the relation

$$q^{\sharp}(\boldsymbol{x})(\boldsymbol{y}) = q(\boldsymbol{x}, \boldsymbol{y}).$$

This map is linear, by the bilinearity of q, and it is one-to-one because q is regular. Since $\dim(E^*) = \dim(E)$, it is an isomorphism. Now consider the composition

$$E \xrightarrow{q^{\sharp}} E^* \xrightarrow{\text{res}} F^*.$$

It is clear that the kernel of this map is $F^{\perp}$, and the claim follows because the restriction map is onto (see 1.2.9), and hence

$$\dim(F^{\perp}) = n - \dim(F^*) = n - \dim(F).$$

(2) It is tautological that $F \subseteq F^{\perp\perp}$, and by (1) both spaces have the same dimension, so they are equal. □

Isotropic and Totally Isotropic Subspaces

The space $F \cap F^{\perp}$ consists of the vectors $\boldsymbol{x} \in F$ that are orthogonal to F. In the Euclidean case, if $F \neq 0$, only 0 can be orthogonal to all vectors of F, and so in this case $F \cap F^{\perp} = \{0\}$ for any F. But in general this is not the case, and in fact it may happen that $F \subseteq F^{\perp}$, as, for example, if $F = \langle \boldsymbol{x} \rangle$, $\boldsymbol{x}$ a null vector.

We say that F is *isotropic* if $F \cap F^{\perp}$ is non-zero. If in addition we have $F \subseteq F^{\perp}$, we say that F is *totally isotropic*. It is clear that this condition is equivalent to say that the restriction of q to F is zero.

1.3.4 *If F is totally isotropic, then* $\dim(F) \leqslant n/2$.

Proof Since $F \subseteq F^{\perp}$, we have $\dim(F) \leqslant \dim(F^{\perp}) = n - \dim(F)$, and therefore $2\dim(F) \leqslant n$. □

1.4 Algebras

We use the term *algebra* to denote a vector space $\mathcal{A}$ (possibly of infinite dimensions) endowed with a bilinear product $\mathcal{A} \times \mathcal{A} \to \mathcal{A}$, $(x, y) \mapsto xy$. In some specific cases, instead of simple juxtaposition of the operands we will use a suitable infix symbol for the product, and in such cases the algebra will be declared explicitly in the form $(\mathcal{A}, *)$, with $*$ denoting the symbol used for the product. If not said otherwise, we also assume that the product is *associative*, which means that

$$(xy)z = x(yz) \;\text{ for all }\; x, y, z \in \mathcal{A},$$

and *unital*, which means that there exists $1_{\mathcal{A}} \in \mathcal{A}$ such that

$$1_{\mathcal{A}} \neq 0_{\mathcal{A}} \text{ and } 1_{\mathcal{A}}x = x1_{\mathcal{A}} = x$$

for all $x \in \mathcal{A}$. The element $1_{\mathcal{A}}$ is unique and is called the *unit* of $\mathcal{A}$.

A linear subspace $\mathcal{B}$ of an algebra $\mathcal{A}$ is said to be a *subalgebra* if $1_{\mathcal{A}} \in \mathcal{B}$ and $bb' \in \mathcal{B}$ for any $b, b' \in \mathcal{B}$. With the induced product bb' it is clearly an algebra with $1_{\mathcal{B}} = 1_{\mathcal{A}}$.

The subalgebra *generated* by a subset S of an algebra $\mathcal{A}$ is the set $\mathbb{R}[S]$ of linear combinations of $1_{\mathcal{A}}$ and arbitrary products of elements of S. If $\mathbb{R}[S] = \mathcal{A}$, we say that S is an *generating set* of $\mathcal{A}$ (as an algebra).

1.4.1 (The Matrix Algebra $\mathbb{R}(n)$) The n^2-dimensional vector space $\mathbb{R}(n)$ of $n \times n$ real matrices is an algebra with the usual product of matrices. Its unit is the identity matrix I_n. □

1.4.2 (The Algebra End(E)) The vector space $\mathrm{End}(E) = \mathrm{Lin}(E, E)$, whose elements are called *endomorphisms* of E, is an algebra with the composition product fg, defined by

$$(fg)(\boldsymbol{x}) = f(g(\boldsymbol{x})).$$

The unit element is the identity $\mathrm{Id} = \mathrm{Id}_E$ of E. By associating to each endomorphism f the matrix $A \in \mathbb{R}(n)$ such that $f(\mathbf{e}) = \mathbf{e}A$ (so the j-th column of A is formed with the components of $f(\boldsymbol{e}_j)$ with respect to $\mathbf{e}$), we get an isomorphism $\mathrm{End}(E) \simeq \mathbb{R}(n)$, which yields $\dim \mathrm{End}(E) = n^2$, in agreement with Eq. (1.3). The main point in the proof is that if B is the matrix of g with respect to $\mathbf{e}$, then

$$(fg)(\mathbf{e}) = f(g(\mathbf{e})) = f(\mathbf{e}B) \overset{(1)}{=} (f\mathbf{e})B = (\mathbf{e}A)B = \mathbf{e}(AB),$$

showing that AB is the matrix of fg. Step (1) is justified by the linearity of f. □

1.4.3 (The Group $\mathcal{A}^\times$) Given any algebra $\mathcal{A}$, let $\mathcal{A}^\times$ be the set of elements $a \in \mathcal{A}$ that have an inverse a^{-1}, which by definition satisfies $aa^{-1} = a^{-1}a = 1_{\mathcal{A}}$. Then it is easily checked that $\mathcal{A}^\times$, with the binary operation induced by the product of $\mathcal{A}$, is a group. This group is called the *multiplicative group* of $\mathcal{A}$ (or also the *group of units* of $\mathcal{A}$). In the case of the algebra End(E) it is clear that $\mathrm{End}(E)^\times = \mathrm{GL}(E)$, the group of linear automorphisms of E, which is called the *general linear group* of E. In terms of matrices, $\mathbb{R}(n)^\times = \mathrm{GL}(n)$, the group of real invertible $n \times n$ matrices.

1.4.4 (Ideals) A linear subspace $\mathcal{I}$ of an algebra $\mathcal{A}$ is called a *left ideal* if $ab \in \mathcal{I}$ for all $a \in \mathcal{A}$ and $b \in \mathcal{I}$. If instead we have $ba \in \mathcal{I}$, we say that $\mathcal{I}$ is a *right ideal.* A *bilateral ideal*, or simply an *ideal*, is a linear subspace $\mathcal{I}$ that is at the same time a left and right ideal, which means that $aba' \in \mathcal{I}$ for all $a, a' \in \mathcal{A}$ and $b \in \mathcal{I}$.

Given a subset S of $\mathcal{A}$, any ideal containing it has to contain all the expressions of the form

$$a_1 s_1 a_1' + \cdots + a_k s_k a_k', \quad a_j, a_j' \in \mathcal{A}, \quad s_j \in S, \quad j = 1, \ldots, k, \quad k \geqslant 0.$$

On the other hand, it is clear that the set of such expressions is an ideal that contains S and hence it is the smallest ideal with this property. We will say that it is the *ideal generated by* S.

1.4.5 (Homomorphisms) If $\mathcal{A}$ and $\mathcal{B}$ are algebras, a linear map $f : \mathcal{A} \to \mathcal{B}$ is called an *algebra homomorphism* if

$$f(aa') = f(a)f(a') \text{ for all } a, a' \in \mathcal{A} \text{ and } f(1_{\mathcal{A}}) = 1_{\mathcal{B}}.$$

The kernel $\ker(f)$ of an algebra homomorphism f is an ideal of $\mathcal{A}$ and its image $f(\mathcal{A})$ is a subalgebra of $\mathcal{B}$.

1.4.6 (Quotient Algebra) This notion is analogous to the notion of quotient of a group by a normal subgroup, and the reasoning used there (see 1.1.6) can be adapted here to show that there exists a quotient and that is unique up to a natural isomorphism. In more detail, a *quotient of an algebra* $\mathcal{A}$ *by an ideal* $\mathcal{I}$ is an onto algebra homomorphism $\pi : \mathcal{A} \to \bar{\mathcal{A}}$ such that $\ker(\pi) = \mathcal{I}$. The main point about such a map, which is called the *universal property of quotients*, is the following: *If* $f : \mathcal{A} \to \mathcal{B}$ *is an algebra homomorphism such that* $\mathcal{I} \subseteq \ker(f)$*, then there exists a unique algebra homomorphism* $\bar{f} : \bar{\mathcal{A}} \to \mathcal{B}$ *such that* $\bar{f}(\pi a) = f(a)$ *for all* $a \in \mathcal{A}$.

Proof First note that if $\pi(a) = \pi(a')$, then $a' = a + z$, with $z \in \mathcal{I}$, and therefore

$$f(a') = f(a + z) = f(a) + f(z) = f(a).$$

This says that the map $\bar{f} : \pi(a) \mapsto f(a)$ is well-defined, for any element $\bar{a} \in \bar{\mathcal{A}}$ can be written in the form $\bar{a} = \pi(a)$, $a \in \mathcal{A}$, and the value of $f(a)$ does not depend on which a we choose. This shows that $\bar{f}$ is uniquely determined, and it is straightforward to check that it is an algebra homomorphism. For instance,

$$\bar{f}(\pi(a)\pi(a')) = \bar{f}(\pi(aa')) = f(aa') = f(a)f(a') = \bar{f}(\pi a)\bar{f}(\pi a').$$

□

If $\pi : \mathcal{A} \to \bar{\mathcal{A}}$ and $\pi' : \mathcal{A} \to \bar{\mathcal{A}}'$ are two quotients of $\mathcal{A}$ by the ideal $\mathcal{I}$, then there is a unique algebra homomorphism $\bar{\pi}' : \bar{\mathcal{A}} \to \bar{\mathcal{A}}'$ such that $\bar{\pi}'(\pi a) = \pi'(a)$ and a unique algebra homomorphism $\bar{\pi} : \bar{\mathcal{A}}' \to \bar{\mathcal{A}}$ such that $\bar{\pi}(\pi' a) = \pi(a)$ (the preceding statement applies to both π' and π). It follows that $\bar{\pi}'\bar{\pi}$ and $\bar{\pi}\bar{\pi}'$ are the identity of $\bar{\mathcal{A}}$ and $\bar{\mathcal{A}}'$, respectively, and hence $\bar{\pi}$ and $\bar{\pi}'$ are inverse isomorphisms. This proves that if there is a quotient, then it is unique up to a natural isomorphism.

The construction of a quotient is easily obtained with the ideas of the proof above. We note that there is bijection of the set $\mathcal{A}/\mathcal{I}$ of subsets of the form $a + \mathcal{I} = \pi^{-1}(\pi a)$, $a \in \mathcal{A}$, and $\bar{\mathcal{A}}$, namely $a + \mathcal{I} \mapsto \pi(a)$. This suggests defining a sum and a product in $\mathcal{A}/\mathcal{I}$ by the rules $(a+\mathcal{I})+(a'+\mathcal{I}) = (a+a')+\mathcal{I}$ and $(a+\mathcal{I})(a'+\mathcal{I}) = (aa')+\mathcal{I}$, and then it is immediate to check that the map $\pi : \mathcal{A} \to \mathcal{A}/\mathcal{I}, a \mapsto a+\mathcal{I}$, is a quotient of $\mathcal{A}$ by $\mathcal{I}$.

Tensor Algebra

To any vector space E it is associated an algebra $(TE, \otimes)$ which is called the *tensor algebra* of E. Its essence is encoded in the following statement:

1.4.7 (Axioms for TE) (1) *E is a linear subspace of TE.* (2) *E generates TE as an algebra and* $1 \notin E$. (3) *If $f : E \to \mathcal{A}$ is a linear map of E into an algebra $\mathcal{A}$, there is an algebra homomorphism $f^{\otimes} : TE \to \mathcal{A}$ such that $f^{\otimes}(\boldsymbol{e}) = f(\boldsymbol{e})$ for all $\boldsymbol{e} \in E$. In particular we have $f^{\otimes}(\boldsymbol{x}_1 \otimes \cdots \otimes \boldsymbol{x}_k) = f(\boldsymbol{x}_1) \cdots f(\boldsymbol{x}_k)$ for any positive integer k and any $\boldsymbol{x}_1, \ldots, \boldsymbol{x}_k \in E$. This relation implies, together with* (2), *that $f^{\otimes}$ is unique.* □

Statement (3) is called the *universal property* of the tensor algebra.

1.4.8 (Example) The goal of this example is to construct an algebra that satisfies the axioms above. Later it will be used to construct a linear basis of TE associated with a basis of E. Note that its construction uses the same idea as the one introduced in the supermarket Example 1.2.2.

Fix a positive integer n and let $N = \{1, \ldots, n\}$. Let $\mathcal{J}$ be the (infinite) set of sequences $J = j_1, \ldots, j_k \in N$. The algebra we are seeking will be based on the vector space $\mathbb{R}[\mathcal{J}]$ of maps $x : \mathcal{J} \to \mathbb{R}$ such that $x(J) = 0$ for all J but a finite number.

For each $J \in \mathcal{J}$, define $\delta_J \in \mathbb{R}[\mathcal{J}]$ by the rules $\delta_J(K) = 0$ if $K \neq J$ and $\delta_J(J) = 1$. The significance of the set $\{\delta_J\}_{J\in\mathcal{J}}$ is that it is a basis of $\mathbb{R}[\mathcal{J}]$. The reason is that for any $x \in \mathbb{R}[\mathcal{J}]$ we obviously have $x = \sum_J x(J)\delta_J$ (notice that this summation makes sense because only a finite number of its terms may be non-zero).

Now define the product $\otimes$ in $\mathbb{R}[\mathcal{J}]$ as the unique bilinear map

$$\mathbb{R}[\mathcal{J}] \times \mathbb{R}[\mathcal{J}] \to \mathbb{R}[\mathcal{J}] \text{ such that } \delta_J \otimes \delta_{J'} = \delta_{J,J'}.$$

Explicitly, this means that $x \otimes x' = \sum_{J,J'\in\mathcal{J}} x(J)x'(J')\,\delta_{J,J'}$. Then $(\mathbb{R}[\mathcal{J}], \otimes)$ is an algebra (its unit is $\delta_{\emptyset}$, and it is associative because the concatenation of sequences is an associative operation), and

$$\delta_{j_1} \otimes \cdots \otimes \delta_{j_k} = \delta_{j_1,\ldots,j_k}. \tag{1.5}$$

If we let $E' = \langle \delta_1, \ldots, \delta_n \rangle$, then E' is a linear subspace of $\mathbb{R}[\mathcal{J}]$ and it is straightforward to check that $(\mathbb{R}[\mathcal{J}], \otimes)$ satisfies the axioms 1.4.7 for E' (see E.1.6, p. 19).

1.4.9 (Functoriality of TE) Let $f : E \to E'$ be a linear map. Then we have a linear map $f : E \to TE'$ by composing f with the inclusion of E' in TE' (no harm in denoting this composition with the same symbol f) and hence there is a unique algebra homomorphism $f^{\otimes} : TE \to TE'$ such that $f^{\otimes}\boldsymbol{x} = f\boldsymbol{x}$ for all $\boldsymbol{x} \in E$. This homomorphism satisfies

$$f^{\otimes}(\boldsymbol{x}_1 \otimes \cdots \otimes \boldsymbol{x}_k) = f(\boldsymbol{x}_1) \otimes \cdots \otimes f(\boldsymbol{x}_k)$$

for any positive integer k and any $\boldsymbol{x}_1, \ldots, \boldsymbol{x}_k \in E$.

In particular, we have the following important special case:

1.4.10 (Parity Involution of TE) There exists a unique algebra automorphism of TE, $x \mapsto \hat{x}$, such that $\hat{\boldsymbol{e}} = -\boldsymbol{e}$ for all $\boldsymbol{e} \in E$. This automorphism is involutive, that is, $\hat{\hat{x}} = x$ for all x, and is called the *parity involution* of TE. □

1.4.11 (Basis of TE Associated with a Basis of E) *Let* $\mathbf{e} = \boldsymbol{e}_1, \ldots, \boldsymbol{e}_n$ *be a basis of* E*. With the notations of* Example 1.4.8, *let* $\boldsymbol{e}_J = \boldsymbol{e}_{j_1} \otimes \cdots \otimes \boldsymbol{e}_{j_k}$, *with the convention that* $\boldsymbol{e}_{\emptyset} = 1$. *Then* $B = \{\boldsymbol{e}_J\}_{J \in \mathcal{J}}$ *is a linear basis of* TE. *In terms of this basis,* $\boldsymbol{e}_J \otimes \boldsymbol{e}_{J'} = \boldsymbol{e}_{J,J'}$.

Proof That B generates TE as a vector space is a consequence of the fact that E generates TE as an algebra and that the tensor product is bilinear. Indeed, expressing each $\boldsymbol{x}_j$ in a product $\boldsymbol{x}_1 \otimes \cdots \otimes \boldsymbol{x}_k$ as a linear combination of $\mathbf{e}$, and expanding using the distributive property of $\otimes$, we get that $\boldsymbol{x}_1 \otimes \cdots \otimes \boldsymbol{x}_k \in \langle B \rangle$, and therefore $\langle B \rangle = TE$.

Now consider the map $E \to E'$ determined by $\boldsymbol{e}_j \mapsto \delta_j$. By functoriality, this map extends to an algebra homomorphism $TE \to \mathbb{R}[\mathcal{J}]$, and so $\boldsymbol{e}_J \mapsto \delta_J$. Since the δ_J are linearly independent, so are the $\boldsymbol{e}_J$. □

1.4.12 (Uniqueness of the Tensor Algebra) The tensor algebra is unique in the following strong sense: if $(TE, \otimes)$ and $(T'E, \otimes')$ satisfy the axioms 1.4.7, then there is a unique algebra isomorphism $TE \simeq T'E$ that extends the identity of E. For the details, see E.1.7, p. 19.

1.4.13 (Existence of the Tensor Algebra) The tensor algebra $(TE, \otimes)$ can be constructed by a modification of the construction of $\mathbb{R}[\mathcal{J}]$ (Example 1.4.8) in which $\mathcal{J}$ is replaced by all finite sequences of vectors. This is the prize to be paid for coming up with a construction that does not depend on having chosen a particular basis. Readers interested in the technicalities are referred to E.1.8, p. 19, for details. It is to be noted, however, that 1.4.11 is sufficient for most of what we need later on. □

Tensor Powers

For any positive integer k, the k-th *tensor power* of E, denoted T^kE, is the linear subspace of TE spanned by all tensor products $\boldsymbol{x}_1 \otimes \cdots \otimes \boldsymbol{x}_k$, $\boldsymbol{x}_1, \ldots, \boldsymbol{x}_k \in E$. Since such products are linear combinations of the set $B_k = \{\boldsymbol{e}_J : |J| = k\}$ (where

$|J|$ denotes the length of the sequence J), we see that $T^k E = \langle B_k \rangle$ and in particular $\dim T^k E = n^k$.

With the convention that $T^0 E = \mathbb{R}$, and on noting that $T^1 E = E$, we get the following decomposition:

$$TE = \bigoplus_{k \geqslant 0} T^k E = \mathbb{R} \oplus E \oplus T^2 E \oplus \cdots,$$

where the $\oplus$ means that any element $x \in TE$ can be written in a unique way as a finite sum $x = \sum_{k \geqslant 0} x_k$ with $x_k \in T^k E$ and $x_k = 0$ for all k but a finite number. Actually if $x = \sum_J \lambda_J \boldsymbol{e}_J$ is the expression of x in terms of the basis B, then $x_k = \sum_{|J|=k} \lambda_J \boldsymbol{e}_J$. Since $T^k E \otimes T^l E \subseteq T^{k+l} E$, we say that the tensor algebra is a *graded algebra*.

1.4.14 (Universal Property of $T^k E$) *If $f : E^k \to E'$ is any k-multilinear map, there exists a unique* linear *map $f^{\otimes k} : T^k E \to E'$ such that*

$$f^{\otimes k}(\boldsymbol{x}_1 \otimes \cdots \otimes \boldsymbol{x}_k) = f(\boldsymbol{x}_1, \ldots, \boldsymbol{x}_k)$$

for all $\boldsymbol{x}_1, \ldots, \boldsymbol{x}_k \in E$.

Proof By 1.2.4 applied to the basis B_k, there exists a unique linear map

$$f^{\otimes k} : T^k E \to E' \text{ such that } f^{\otimes k}(\boldsymbol{e}_J) = f(\boldsymbol{e}_{j_1}, \ldots, \boldsymbol{e}_{j_k}).$$

Now the expression $f^{\otimes k}(\boldsymbol{x}_1 \otimes \cdots \otimes \boldsymbol{x}_k)$ yields a k-linear map $E^k \to E'$ and by 1.2.4′ this map agrees with f. □

1.4.15 (Remark on the Parity Involution) For the parity involution of TE we have the formula $\hat{x} = (-1)^k x$ for all $x \in T^k E$. This is clear from the fact that

$$(\boldsymbol{x}_1 \otimes \cdots \otimes \boldsymbol{x}_k)\hat{} = \hat{\boldsymbol{x}}_1 \otimes \cdots \otimes \hat{\boldsymbol{x}}_k = (-\boldsymbol{x}_1) \otimes \cdots \otimes (-\boldsymbol{x}_k) = (-1)^k \boldsymbol{x}_1 \otimes \cdots \otimes \boldsymbol{x}_k.$$

1.4.16 (Reverse Involution of TE) There is also a unique algebra *antiautomorphism* of TE, $x \mapsto \tilde{x}$, such that $(\boldsymbol{x}_1 \otimes \cdots \otimes \boldsymbol{x}_k)^{\sim} = \boldsymbol{x}_k \otimes \cdots \otimes \boldsymbol{x}_1$ for all $k \geqslant 1$ and $\boldsymbol{x}_1, \ldots, \boldsymbol{x}_k \in E$. It is clearly involutive and it is called the *reverse involution* of TE.

Proof Consider the linear map $E \to T^{\mathrm{op}} E$, where $T^{\mathrm{op}} E$ is the same space as TE, but with the product $x \otimes^{\mathrm{op}} x' = x' \otimes x$. Then there is a unique algebra homomorphism $TE \to T^{\mathrm{op}} E$ extending the identity of E, and this homomorphism satisfies $(\boldsymbol{x}_1, \ldots, \boldsymbol{x}_k) \mapsto \boldsymbol{x}_k \otimes \cdots \otimes \boldsymbol{x}_1$. □

1.4.17 (Contraction Operators) Fix a linear form $\boldsymbol{\xi} \in E^*$ and consider, for $k \geqslant 1$, the map $E^k \to T^{k-1} E$ defined by

$$(\boldsymbol{y}_1, \ldots, \boldsymbol{y}_k) \mapsto \sum_{j=1}^{k} (-1)^{j-1} \boldsymbol{\xi}(\boldsymbol{y}_j)\, \boldsymbol{y}_1 \otimes \cdots \otimes \boldsymbol{y}_{j-1} \otimes \boldsymbol{y}_{j+1} \otimes \cdots \otimes \boldsymbol{y}_k.$$

Since this map is clearly k-linear, there exists a unique linear map $i_{\xi} : T^k E \to T^{k-1} E$ such that

$$i_{\xi}(\boldsymbol{y}_1 \otimes \cdots \otimes \boldsymbol{y}_k) = \sum_{j=1}^{k} (-1)^{j-1} \xi(\boldsymbol{y}_j)\, \boldsymbol{y}_1 \otimes \cdots \otimes \boldsymbol{y}_{j-1} \otimes \boldsymbol{y}_{j+1} \otimes \cdots \otimes \boldsymbol{y}_k. \quad (1.6)$$

With the natural convention that $i_{\xi}(\lambda) = 0$ for any scalar λ (natural on account that $T^0 E = \mathbb{R}$ and $T^{-1} E = \{0\}$), we get a linear map $i_{\xi} : TE \to TE$ of grade -1. This map, which is called the ξ-*contraction operator*, is a *skew-derivation*, which means that

$$i_{\xi}(x \otimes y) = (i_{\xi} x) \otimes y + \hat{x} \otimes (i_{\xi} y) \quad (1.7)$$

for all $x, y \in TE$. To prove this it is enough to check the case in which x and y are tensor products of vectors, which in turn is an immediate consequence of Eq. (1.6).

If E is endowed with a metric q, then for any $\boldsymbol{x} \in E$ we have the linear form $q_{\boldsymbol{x}}$ defined by $q_{\boldsymbol{x}}(\boldsymbol{y}) = q(\boldsymbol{x}, \boldsymbol{y})$. The $\boldsymbol{x}$-*contraction operator* $i_{\boldsymbol{x}}$ is defined by $i_{\boldsymbol{x}} = i_{q_{\boldsymbol{x}}}$. It is thus clear that $i_{\boldsymbol{x}}$ is a grade -1 skew-derivation of TE which satisfies

$$i_{\boldsymbol{x}}(\boldsymbol{y}_1 \otimes \cdots \otimes \boldsymbol{y}_k) = \sum_{j=1}^{k} (-1)^{j-1} q(\boldsymbol{x}, \boldsymbol{y}_j) \boldsymbol{y}_1 \otimes \cdots \otimes \boldsymbol{y}_{j-1} \otimes \boldsymbol{y}_{j+1} \otimes \cdots \otimes \boldsymbol{y}_k. \quad (1.8)$$

In fact it is an easy exercise to see that i_{ξ} ($i_{\boldsymbol{x}}$) is the unique skew-derivation of TE such that $i_{\xi}(\boldsymbol{y}) = \xi(\boldsymbol{y})$ ($i_{\xi}(\boldsymbol{y}) = q(\boldsymbol{x}, \boldsymbol{y})$) for any vector $\boldsymbol{y}$. Notice that $i_{\xi}(\lambda) = 0$ for any scalar λ follows from $i_{\xi}(\lambda) = i_{\xi}(\lambda 1) = i_{\xi}(\lambda)1 + \lambda i_{\xi}(1)$.

1.5 Exercises

E.1.1 If $\pi : G \to \bar{G}$ is a quotient of G, then G is finite if and only if $H = \ker(\pi)$ and $\bar{G}$ are finite, and if this is the case, then $|G| = |\bar{G}||H|$.

Hint We have that $|G| = \sum_{\bar{x} \in \bar{G}} |\pi^{-1}\bar{x}|$, and $\pi^{-1}\bar{x} = xH$ for any $x \in G$ such that $\pi(x) = \bar{x}$, so that $|\pi^{-1}\bar{x}| = |H|$ for any $\bar{x} \in \bar{G}$.

E.1.2 Let F be a vector subspace of the vector space E, and $\boldsymbol{\beta} : F \to \mathbb{R}$ a linear map. Show that there exists a linear map $\boldsymbol{\alpha} : E \to \mathbb{R}$ such that $\boldsymbol{\alpha}(\boldsymbol{x}) = \boldsymbol{\beta}(\boldsymbol{x})$ for all $\boldsymbol{x} \in F$.

Hint Take a linear subspace G of E such that $E = F + G$ and $F \cap G = \{0\}$. Then any $\boldsymbol{x} \in E$ can be decomposed in a unique way as $\boldsymbol{x} = \boldsymbol{y} + \boldsymbol{z}$ with $\boldsymbol{y} \in F$ and $\boldsymbol{z} \in G$. Define $\boldsymbol{\alpha}$ by $\boldsymbol{\alpha}(\boldsymbol{x}) = \boldsymbol{\beta}(\boldsymbol{y})$.

E.1.3 (Polarization Formula) Prove the polarization formula (1.4).

E.1.4 (Existence of Orthogonal Basis) Let q be an arbitrary symmetric bilinear form of a vector space E, and let $n = \dim(E)$. Then there exist q-orthogonal bases of E.

Hint If all vectors are isotropic, then $q = 0$ (by the polarization formula) and any basis is q-orthogonal. So we may assume that there is a non-isotropic vector $\boldsymbol{e}_1$. Since the linear map $E \to \mathbb{R}$, $\boldsymbol{x} \mapsto q(\boldsymbol{e}_1, \boldsymbol{x})$ does not vanish at $\boldsymbol{e}_1$, the map is surjective, $E' = \{\boldsymbol{x} \in E : q(\boldsymbol{e}_1, \boldsymbol{x}) = 0\}$ has dimension $n-1$, and $E = \langle \boldsymbol{e}_1 \rangle + E'$. Now by induction there exists q-orthogonal basis $\boldsymbol{e}_2, \ldots, \boldsymbol{e}_n$ of E' and then $\boldsymbol{e}_1, \boldsymbol{e}_2, \ldots, \boldsymbol{e}_n$ is an q-orthogonal basis of E.

E.1.5 (Sylvester's Law) Let q be a metric of the vector space E. Let $\boldsymbol{e}_1, \ldots, \boldsymbol{e}_n$ be a q-orthogonal basis of E. The signature of this basis is the pair (r, s) of non-negative integers giving the number r of indices j such that $q(\boldsymbol{e}_j) > 0$ and the number s of indices j such that $q(\boldsymbol{e}_j) < 0$). Prove that any two q-orthogonal bases have the same signature.

Hint Suppose that $\boldsymbol{e}'_1, \ldots, \boldsymbol{e}'_m$ are positive pairwise orthogonal vectors. Reindexing the basis so that $\boldsymbol{e}_1, \ldots, \boldsymbol{e}_r$ are positive, consider the map linear map $F = \langle \boldsymbol{e}'_1, \ldots, \boldsymbol{e}'_m \rangle \to \mathbb{R}^r$, $\boldsymbol{x} \mapsto (q(\boldsymbol{x}, \boldsymbol{e}_1), \ldots, q(\boldsymbol{x}, \boldsymbol{e}_r))$. If a vector $\boldsymbol{x} \in F$ maps to 0, it is a linear combination of $\boldsymbol{e}_{r+1}, \ldots, \boldsymbol{e}_n$ and hence $q(\boldsymbol{x}) \leqslant 0$. But we also have $q(\boldsymbol{x}) \geqslant 0$, so $q(\boldsymbol{x}) = 0$. This implies that $\boldsymbol{x} = 0$ and we can conclude that $m \leqslant r$.

E.1.6 (The Algebra $\mathbb{R}[\mathcal{J}]$) Following the notations setup in Example 1.4.8, we want to show that $(\mathbb{R}[\mathcal{J}], \otimes)$ satisfies the axioms 1.4.7 for $E' = \langle \delta_1, \ldots, \delta_n \rangle$. Since the properties (1) and (2) are clear (recall Eq. (1.5)), it only remains to prove (3).

Let $f : E' \to \mathcal{A}$ be a linear map from E' into an algebra $\mathcal{A}$. If for any $J \in \mathcal{J}$ we set $e'_J = f(\delta_{j_1}) \cdots f(\delta_{j_k})$, then there is unique linear map $f^* : \mathbb{R}[\mathcal{J}] \to \mathcal{A}$ such that $\delta_J \mapsto e'_J$ (this follows from a straightforward extension of the principle 1.2.4). This map satisfies

$$f^*(\delta_J \otimes \delta_{J'}) = f^*(\delta_{J,J'}) = f(\delta_{j_1}) \cdots f(\delta_{j_k}) f(\delta_{j'_1}) \cdots f(\delta_{j'_{k'}}) = f^*(\delta_J) f^*(\delta_{J'})$$

and this implies that f^* is an algebra homomorphism extending f.

E.1.7 (Uniqueness of the Tensor Algebra) If $(TE, \otimes)$ and $(T'E, \otimes')$ satisfy the conditions of 1.4.7, then the inclusion of E in $T'E$ can be extended uniquely to an algebra homomorphism $TE \to T'E$, and this homomorphism satisfies

$$\boldsymbol{x}_1 \otimes \cdots \otimes \boldsymbol{x}_k \mapsto \boldsymbol{x}_1 \otimes' \cdots \otimes' \boldsymbol{x}_k$$

for any positive integer k and any $\boldsymbol{x}_1, \ldots, \boldsymbol{x}_k \in E$. *This homomorphism is an isomorphism*, for there is also, for the same reasons, an algebra homomorphism $T'E \to TE$ such that $\boldsymbol{x}_1 \otimes' \cdots \otimes' \boldsymbol{x}_k \mapsto \boldsymbol{x}_1 \otimes \cdots \otimes \boldsymbol{x}_k$, and then the compositions $TE \to T'E \to TE$ and $T'E \to TE \to T'E$ are the identity because they satisfy $\boldsymbol{x}_1 \otimes \cdots \otimes \boldsymbol{x}_k \mapsto \boldsymbol{x}_1 \otimes \cdots \otimes \boldsymbol{x}_k$ and $\boldsymbol{x}_1 \otimes' \cdots \otimes' \boldsymbol{x}_k \mapsto \boldsymbol{x}_1 \otimes' \cdots \otimes' \boldsymbol{x}_k$, respectively.

E.1.8 (Existence of the Tensor Algebra) Let E be a vector space of finite dimension n. Let Σ be the set of all finite sequences of elements of E and consider the vector space $\mathbb{R}[\Sigma]$ of all maps $x : \Sigma \to \mathbb{R}$ such that $x(s) = 0$ except for a finite number of s. For each $s \in \Sigma$, let $\epsilon_s \in \mathbb{R}[\Sigma]$ be the map such that $\epsilon_s(s) = 1$

and $\epsilon_s(t) = 0$ for all $t \neq s$. Then $B = \{\epsilon_s : s \in \Sigma\}$ is a basis of $\mathbb{R}[\Sigma]$, for $x = \sum x(s)\epsilon_s$ for any $x \in \mathbb{R}[\Sigma]$ (the summation involves only a finite number of non-zero terms). Now we can turn $\mathbb{R}[\Sigma]$ into an associative algebra by defining the product $\boxtimes : \mathbb{R}[\Sigma] \times \mathbb{R}[\Sigma] \to \mathbb{R}[\Sigma]$ as the unique bilinear map such that $\epsilon_s \boxtimes \epsilon_{s'} = \epsilon_{s,s'}$. Its unit is $1 = \epsilon_\emptyset$. Now let $\mathcal{I}$ be the ideal of $\mathbb{R}[\Sigma]$ generated by the elements of the form $\epsilon_{\boldsymbol{x}+\boldsymbol{y}} - \epsilon_{\boldsymbol{x}} - \epsilon_{\boldsymbol{y}}$ and $\epsilon_{\lambda\boldsymbol{x}} - \lambda\epsilon_{\boldsymbol{x}}$, where $\boldsymbol{x}, \boldsymbol{y} \in E$ and $\lambda \in \mathbb{R}$ (note that vectors are vector sequences of length 1). Finally let $(TE, \otimes)$ be the quotient of $\mathbb{R}[\Sigma]/\mathcal{I}$.

There is a natural set inclusion $E \to \mathbb{R}[\Sigma]$, $\boldsymbol{x} \mapsto \epsilon_{\boldsymbol{x}}$. If we write $\bar{\boldsymbol{x}}$ to denote the image of $\epsilon_{\boldsymbol{x}}$ in the quotient TE, then the natural map $E \to TE$, $\boldsymbol{x} \mapsto \bar{\boldsymbol{x}}$, is *linear*: $\overline{\boldsymbol{x}+\boldsymbol{y}} = \bar{\boldsymbol{x}} + \bar{\boldsymbol{y}}$ and $\overline{\lambda\boldsymbol{x}} = \lambda\bar{\boldsymbol{x}}$.

Now let $f : E \to \mathcal{A}$ be a linear map of E into an algebra $\mathcal{A}$. Then f extends to a map $f^{\boxtimes} : \Sigma \to \mathcal{A}$ in a natural way, namely $f^{\boxtimes}(\boldsymbol{x}_1, \ldots, \boldsymbol{x}_k) = f(\boldsymbol{x}_1) \cdots f(\boldsymbol{x}_k)$, and hence to a linear map $f^{\boxtimes} : \mathbb{R}[\Sigma] \to \mathcal{A}$ (no harm in using the same symbol) such that $\epsilon_s \mapsto f^{\boxtimes}(s)$. From the definitions, it follows that $f^{\boxtimes}$ is actually an algebra homomorphism; and from the fact that $\mathcal{A}$ is an algebra, it follows that $\mathcal{I}$ is included in the kernel of $f^{\boxtimes}$. Therefore we get an algebra homomorphism $f^{\otimes} : TE \to \mathcal{A}$, that extends f.

E.1.9 (Tensor Product of Spaces and Algebras) A *tensor product* of two vector spaces E and E' is a vector space T and a bilinear map $t : E \times E' \to T$ that has the following property: for any bilinear map $f : E \times E' \to F$, there exists a unique *linear* map $\bar{f} : T \to F$ such that $\bar{f}(t(\boldsymbol{x}, \boldsymbol{x}')) = f(\boldsymbol{x}, \boldsymbol{x}')$ for all $\boldsymbol{x} \in E, \boldsymbol{x}' \in E'$. For example, to the bilinear map t there corresponds a unique linear map $\bar{t} : T \to T$ such that $\bar{t}(t(\boldsymbol{x}, \boldsymbol{x}')) = t(\boldsymbol{x}, \boldsymbol{x}')$, and uniqueness implies that $\bar{t} = \mathrm{Id}_T$. Similarly, if $s : E \times E' \to S$ is another tensor product, then we have a unique linear map $\bar{s} : T \to S$ such that $\bar{s}(t(\boldsymbol{x}, \boldsymbol{x}')) = s(\boldsymbol{x}, \boldsymbol{x}')$, a unique linear map $\bar{t} : S \to T$ such that $\bar{t}(s(\boldsymbol{x}, \boldsymbol{x}')) = t(\boldsymbol{x}, \boldsymbol{x}')$, and both $\bar{s} \circ \bar{t} : S \to S$ and $\bar{t} \circ \bar{s} : T \to T$ must be the identity. Therefore the tensor product, it exists, is unique up to a unique linear isomorphism. Its existence can be proved by a straightforward adaptation of the existence argument in the preceding exercise.

The tensor product is denoted $E \otimes E'$ and instead of $t(\boldsymbol{x}, \boldsymbol{x}')$ we write $\boldsymbol{x} \otimes \boldsymbol{x}'$. Restating the defining property with these notations, we have that for any bilinear map $f : E \times E' \to F$ there exists a unique linear map $f^{\otimes} : E \otimes E' \to F$ such that $f^{\otimes}(\boldsymbol{x} \otimes \boldsymbol{x}') = f(\boldsymbol{x}, \boldsymbol{x}')$.

The tensor product of a finite number of vector spaces $E_1 \otimes \cdots \otimes E_k$ is defined in a similar way. At the end we have a k-linear map $E_1 \times \cdots \times E_k \to E_1 \otimes \cdots \otimes E_k$ such that for any k-linear map $f : E_1 \times \cdots \times E_k \to F$ there exists a unique linear map $f^{\otimes} : E_1 \otimes \cdots \otimes E_k \to F$ such that $f^{\otimes}(\boldsymbol{x}_1 \otimes \cdots \otimes \boldsymbol{x}_k) = f(\boldsymbol{x}_1, \ldots, \boldsymbol{x}_k)$. Observe that $T^k E \simeq E \otimes \stackrel{k)}{\cdots} \otimes E$.

If $\mathcal{A}$ and $\mathcal{A}'$ are algebras, then the space $\mathcal{A} \otimes \mathcal{A}'$ has a unique bilinear product $(\mathcal{A} \otimes \mathcal{A}') \times (\mathcal{A} \otimes \mathcal{A}') \to \mathcal{A} \otimes \mathcal{A}'$ such that $(a \otimes a', b \otimes b') \mapsto ab \otimes a'b'$. This endows $\mathcal{A} \otimes \mathcal{A}'$ with a bilinear product which is associative if $\mathcal{A}$ and $\mathcal{A}'$ are associative. If $\mathcal{A}$ and $\mathcal{A}'$ are unital, then $\mathcal{A} \otimes \mathcal{A}'$ is unital and its unit is $1_{\mathcal{A}} \otimes 1_{\mathcal{A}'}$.

E.1.10 (Examples)

(1) $\mathbb{R} \otimes E = E \otimes \mathbb{R} = E$, with $\lambda \otimes \boldsymbol{x} = \boldsymbol{x} \otimes \lambda = \lambda \boldsymbol{x}$. In fact, if $f : \mathbb{R} \times E \to F$ is bilinear, the map $\bar{f} : E \to F, \boldsymbol{x} \mapsto f(1, \boldsymbol{x})$ *is linear* and $\bar{f}(\lambda \otimes \boldsymbol{x}) = \bar{f}(\lambda \boldsymbol{x}) = f(1, \lambda \boldsymbol{x}) = f(\lambda, \boldsymbol{x})$, again because f is bilinear. This is also true for algebras: $\mathbb{R} \otimes \mathcal{A} = \mathcal{A} \otimes \mathbb{R} = \mathcal{A}$ as algebras.

(2) If $\mathcal{A}$ is an algebra, and m a positive integer, $\mathcal{A} \otimes \mathbb{R}(m) = \mathcal{A}(m)$, with $a \otimes M = aM$ ($a \in \mathcal{A}$, $M \in \mathbb{R}(m)$, and the obvious definition of aM). In particular,

$$\mathbb{R}(m) \otimes \mathbb{R}(m') = \mathbb{R}(mm'). \tag{1.9}$$

(3) A last example is that

$$\mathbb{C} \otimes \mathbb{C} = \mathbb{C} \oplus \mathbb{C}. \tag{1.10}$$

Indeed, since $(i \otimes i)^2 = i^2 \otimes i^2 = 1 \otimes 1$, the elements $e_\pm = \frac{1}{2}(1 \otimes 1 \pm i \otimes i)$ satisfy $e_+ + e_- = 1 \otimes 1$, $e_\pm^2 = e_\pm$, and $e_+ e_- = e_- e_+ = 0 \otimes 0$. As a consequence, the linear map $\mathbb{C} \oplus \mathbb{C} \to \mathbb{C} \otimes \mathbb{C}$, $(z, w) \mapsto ze_+ + wz_-$, is an algebra isomorphism, because it is a linear isomorphism that satisfies

$$(ze_+ + we_-)(z'e_+ + w'e_-) = zz'e_+ + ww'e_-.$$

Chapter 2
Grassmann Algebra

After presenting the exterior algebra in Sect. 2.1, the remaining three sections are devoted to the main aspects of this algebra that depend on the metric, namely the contraction operator, Sect. 2.2, the extension of the metric to the whole algebra, Sect. 2.3, and the inner product, Sect. 2.4. The point of view here is to call *Grassmann algebra* to the structure formed by the exterior algebra enriched with the metric, the inner product, and the parity and reverse involutions.

This approach may seem a distraction detour before studying geometric algebras, but we will see that this is not the case. Some key properties of geometric algebra (in particular the existence of a canonical linear grading) presuppose the knowledge of the exterior algebra, and in any case all the ingredients of the exterior algebra can be neatly grafted into the geometric algebra and readily used for the study and handling of this richer structure.

2.1 Exterior Algebra

The exterior algebra is the algebraic structure discovered by H. Grassmann in his *Ausdehnungslehre* [58, 59] (Extension Theory, [60]). Besides vectors, which in its original context were thought as oriented segments, Grassmann conceived higher-dimensional analogues, like oriented areas, oriented volumes, and so on, and discovered the algebraic laws that govern them. For example, given two vectors $\boldsymbol{x}$ and $\boldsymbol{x}'$, there is a two-dimensional oriented extension, which is denoted by $\boldsymbol{x} \wedge \boldsymbol{x}'$, associated to the parallelogram they define (see Fig. 2.1). The fact that the swapping of the vectors reverses the orientation of the parallelogram is translated into the algebraic law $\boldsymbol{x}' \wedge \boldsymbol{x} = -\boldsymbol{x} \wedge \boldsymbol{x}'$. In particular, we have $\boldsymbol{x} \wedge \boldsymbol{x} = 0$, which corresponds to the fact that the parallelogram defined by $\boldsymbol{x}$ and $\boldsymbol{x}$ represents a null two-dimensional extension.

S. Xambó-Descamps, *Real Spinorial Groups*, SpringerBriefs in Mathematics,
https://doi.org/10.1007/978-3-030-00404-0_2

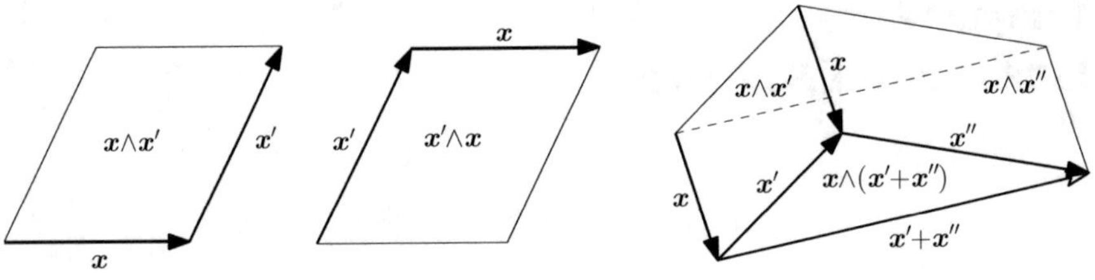

Fig. 2.1 Properties of the exterior product of vectors: $\boldsymbol{x}' \wedge \boldsymbol{x} = -\boldsymbol{x} \wedge \boldsymbol{x}'$ (left and middle) and $\boldsymbol{x} \wedge (\boldsymbol{x}' + \boldsymbol{x}'') = \boldsymbol{x} \wedge \boldsymbol{x}' + \boldsymbol{x} \wedge \boldsymbol{x}''$ (right)

Presently, the most expedient way to discuss Grassmann's ideas is to first set up the formal algebraic structure and elicit the geometric interpretations afterwards.

To any vector space E, there is associated an algebra $(\wedge E, \wedge)$, called the *exterior algebra* of E. Its structural formal features can be encapsulated as follows:

2.1.1 (Axioms for $\wedge E$) (1) E is a linear subspace of $\wedge E$ and $\boldsymbol{x} \wedge \boldsymbol{x} = \boldsymbol{x}^{\wedge 2} = 0$ for all $\boldsymbol{x} \in E$. (2) E generates $\wedge E$ as an algebra and $1 \notin E$. (3) If $f : E \to \mathcal{A}$ is a linear map of E into an algebra $(\mathcal{A}, *)$ such that $(f\boldsymbol{x})^{*2} = 0$ for all $\boldsymbol{x} \in E$, then there is an algebra homomorphism $f^{\wedge} : \wedge E \to \mathcal{A}$ such that $f^{\wedge}(\boldsymbol{x}) = f(\boldsymbol{x})$ for all $\boldsymbol{x} \in E$. In particular, we have that

$$f^{\wedge}(\boldsymbol{x}_1 \wedge \cdots \wedge \boldsymbol{x}_k) = f(\boldsymbol{x}_1) * \cdots * f(\boldsymbol{x}_k)$$

for any positive integer k and any $\boldsymbol{x}_1, \ldots, \boldsymbol{x}_k \in E$. This relation implies, together with (2), that $f^{\wedge}$ is unique. □

2.1.2 (Remark) The condition $\boldsymbol{x} \wedge \boldsymbol{x} = 0$ for all $\boldsymbol{x} \in E$ is equivalent to say that $\boldsymbol{x} \wedge \boldsymbol{y} = -\boldsymbol{y} \wedge \boldsymbol{x}$ for all $\boldsymbol{x}, \boldsymbol{y} \in E$ (expand $0 = (\boldsymbol{x}+\boldsymbol{y}) \wedge (\boldsymbol{x}+\boldsymbol{y})$ to get $\boldsymbol{x} \wedge \boldsymbol{y} + \boldsymbol{y} \wedge \boldsymbol{x} = 0$). □

2.1.3 (Linear Generators of $\wedge E$ Associated to a Basis of E) *Let* $\mathbf{e} = \boldsymbol{e}_1, \ldots, \boldsymbol{e}_n$ *be a basis of E. For any sequence $J = j_1, \ldots, j_k \in N = \{1, \ldots, n\}$, let $\boldsymbol{e}_{\hat{J}} = \boldsymbol{e}_{j_1} \wedge \cdots \wedge \boldsymbol{e}_{j_k}$, with the convention that $\boldsymbol{e}_{\hat{\emptyset}} = 1$. Then, $\wedge E$ is the linear span of $\hat{B} = \{\boldsymbol{e}_{\hat{J}}\}_{J \in \hat{\mathcal{J}}}$, where $\hat{\mathcal{J}}$ denotes the set of all* strictly increasing *sequences of* N (which will also be called multiindices), *and the table of exterior products of elements of $\hat{B}$ goes as follows:*

$$\boldsymbol{e}_{\hat{J}} \wedge \boldsymbol{e}_{\hat{K}} = \begin{cases} 0 & \text{if } J \cap K \neq \emptyset \\ (-1)^{t(J,K)} \boldsymbol{e}_{\hat{L}} & \text{if } J \cap K = \emptyset \end{cases}$$

where L is the result of arranging $J \cup K$ in increasing order.

Proof The exterior products of the form $\boldsymbol{x}_1 \wedge \cdots \wedge \boldsymbol{x}_k$ ($k \geqslant 1, \boldsymbol{x}_1, \ldots, \boldsymbol{x}_k \in E$), together with 1, are linear generators of $\wedge E$. Since $\boldsymbol{x}_1 \wedge \cdots \wedge \boldsymbol{x}_k$ is a k-linear expression, we find that the $\boldsymbol{e}_{\hat{J}}$, $J \in \mathcal{J}$, are also linear generators of $\wedge E$. Since $\boldsymbol{e}_{\hat{J}}$

has the same value, up to sign, for all permutations of J, and since it vanishes if J has repeated indices, we get that also the $\boldsymbol{e}_{\hat{J}}$, $J \in \hat{\mathcal{J}}$, are linear generators of $\wedge E$.

The product $\boldsymbol{e}_{\hat{J}} \wedge \boldsymbol{e}_{\hat{K}}$ is clearly equal to $(-1)^{t(J,K)}\boldsymbol{e}_{\hat{L}}$, where L is the result of rearranging the sequence J, K in non-decreasing order. If $J \cap K \neq \emptyset$, L has a repeated index for any $i \in J \cap K$, and any such repetition contributes to $\boldsymbol{e}_{\hat{L}}$ as $\boldsymbol{e}_i \wedge \boldsymbol{e}_i = 0$. □

2.1.4 (Example) This example is the analog for the exterior algebra of Example 1.4.8 for the tensor algebra. Let $\mathbb{R}[\hat{\mathcal{J}}]$ be the vector space of maps $x : \hat{\mathcal{J}} \to \mathbb{R}$ (no restrictions on x are needed because $\hat{B}$ is finite). Then, the set of the $\delta_J \in \mathbb{R}[\hat{\mathcal{J}}]$ ($J \in \hat{\mathcal{J}}$) defined by $\delta_J(K) = 0$ if $K \neq J$, and $\delta_J(J) = 1$, is a basis of $\mathbb{R}[\hat{\mathcal{J}}]$, as for any $x \in \mathbb{R}[\hat{\mathcal{J}}]$, we have $x = \sum_{J\in\hat{\mathcal{J}}} x(J)\delta_J$.

It is convenient to define $\delta_J \in \mathbb{R}[\hat{\mathcal{J}}]$ for any index sequence J as follows: $\delta_J = 0$ if J has repeated indices, and otherwise $\delta_J = (-1)^{t(J)}\delta_{\tilde{J}}$, where $\tilde{J}$ is the result of rearranging J in increasing order.

Define the product $\wedge$ in $\mathbb{R}[\hat{\mathcal{J}}]$ as the unique bilinear map:

$$\mathbb{R}[\hat{\mathcal{J}}] \times \mathbb{R}[\hat{\mathcal{J}}] \to \mathbb{R}[\hat{\mathcal{J}}] \ \text{ such that } \ \delta_J \wedge \delta_K = \delta_{J,K}.$$

In other words, $\delta_J \wedge \delta_K = 0$ if $J \cap K \neq \emptyset$ and $\delta_J \wedge \delta_K = (-1)^{t(J,K)}\delta_{(J,K)^{\sim}}$ if $J \cap K = \emptyset$. Then, it is straightforward to check that $(\mathbb{R}[\hat{\mathcal{J}}], \wedge)$ is an algebra (for the associativity, see E.2.1, p. 37) with unit $\delta_\emptyset$, and that for $J \in \hat{\mathcal{J}}$,

$$\delta_{j_1} \wedge \cdots \wedge \delta_{j_k} = \delta_{j_1,\ldots,j_k}. \tag{2.1}$$

Now, if we set $E' = \langle \delta_1, \ldots, \delta_n \rangle$, then E' is a linear subspace of $\mathbb{R}[\hat{\mathcal{J}}]$, and it is easy to see that $(\mathbb{R}[\hat{\mathcal{J}}], \wedge)$ satisfies the axioms 2.1.1 for E' (see E.2.2, p. 37).

2.1.5 (Functoriality of $\wedge E$) Let $f : E \to E'$ be a linear map. Then, we have a linear map $f : E \to \wedge E'$ by composing f with the inclusion of E' in $\wedge E'$ (no harm in denoting this composition with the same symbol f). Since $(f\boldsymbol{x})^{\wedge 2} = 0$ for all $\boldsymbol{x} \in E$ is satisfied, there is a unique algebra homomorphism $f^\wedge : \wedge E \to \wedge E'$ such that $f^\wedge(\boldsymbol{x}) = f(\boldsymbol{x})$ for all $\boldsymbol{x} \in E$. This homomorphism satisfies

$$f^\wedge(\boldsymbol{x}_1 \wedge \cdots \wedge \boldsymbol{x}_k) = f(\boldsymbol{x}_1) \wedge \cdots \wedge f(\boldsymbol{x}_k) \tag{2.2}$$

for any positive integer k and any $\boldsymbol{x}_1, \ldots, \boldsymbol{x}_k \in E$. □

2.1.6 (Basis of $\wedge E$ Associated to a Basis of E) *The set $\hat{B}$ introduced in* 2.1.3 *is linearly independent. In particular,* $\dim \wedge E = 2^n$.

Proof We will use the notations and results of Example 2.1.4. Given the basis $\boldsymbol{e}_1, \ldots, \boldsymbol{e}_n$ of E, there is a unique linear map $E \to E'$ such that $\boldsymbol{e}_j \mapsto \delta_j$ and by functoriality this map extends to an algebra homomorphism $\wedge E \to \mathbb{R}[\hat{\mathcal{J}}]$ such that $\boldsymbol{e}_{\hat{J}} \mapsto \delta_J$. Since the δ_J are linearly independent, so are the $\boldsymbol{e}_{\hat{J}}$. □

2.1.7 (Uniqueness of $\wedge E$) The exterior algebra is unique in the following strong sense: if $(\wedge E, \wedge)$ and $(\wedge' E, \wedge')$ satisfy the axioms 2.1.5, then there is a unique algebra isomorphism $\wedge E \simeq \wedge' E$ that extends the identity of E. For the details, which mimic those of the uniqueness of the tensor algebra, see E.2.3, p. 37. □

2.1.8 (Existence of $\wedge E$) In Example 2.1.4, we have constructed a particular exterior algebra, but its bearing on $\wedge E$, as in 2.1.6, depends on choosing a basis of E. To produce a general construction, instead of adapting the construction of the tensor algebra outlined in E.1.8, p. 19 (which is definitely possible), it is better to use the existence of the tensor algebra. In fact, given a linear map $f : E \to \mathcal{A}$ of E into an algebra $(\mathcal{A}, *)$ such that $(f\boldsymbol{x})^{*2} = 0$ for all $\boldsymbol{x} \in E$, we can consider its unique extension to an algebra homomorphism $f^{\otimes} : TE \to \mathcal{A}$, and this homomorphism has the property that

$$f^{\otimes}(\boldsymbol{x} \otimes \boldsymbol{x}) = (f\boldsymbol{x}) * (f\boldsymbol{x}) = 0.$$

This says that if we let $\mathcal{I}$ denote the ideal of TE generated by the elements of the form $\boldsymbol{x} \otimes \boldsymbol{x}$, $\boldsymbol{x} \in E$, then we have an algebra homomorphism $TE/\mathcal{I} \to \mathcal{A}$ that extends f (in the sense that the composition $E \to TE \to TE/\mathcal{I} \to \mathcal{A}$ coincides with f). And now, it is straightforward to check that if we define $\wedge E = TE/\mathcal{I}$, and let $\wedge$ stand for the product of $\wedge E$ induced by $\otimes$, then $(\wedge E, \wedge)$ satisfies the axioms 2.1.1. For the details, see E.2.4, p. 38.

Exterior Powers

The k-th *exterior power* of E ($k \geqslant 1$), which is denoted by $\wedge^k E$, is the linear subspace of $\wedge E$ generated by the image of the k-linear alternating map $E^k \to \wedge E$, $(\boldsymbol{x}_1, \ldots, \boldsymbol{x}_k) \mapsto \boldsymbol{x}_1 \wedge \cdots \wedge \boldsymbol{x}_k$. Conventionally, we set $\wedge^0 E = \mathbb{R}$. With the notations of 2.1.1, a basis for this space is the set $\hat{B}_k = \{\boldsymbol{e}_{\hat{J}}\}_{|J|=k}$, so that $\dim \wedge^k E = \binom{n}{k}$. Note that $\wedge^k E = 0$ for $k > n$.

Then, we clearly have

$$\wedge E = \bigoplus_{k=0}^{n} \wedge^k E = \mathbb{R} \oplus E \oplus \wedge^2 E \oplus \cdots \oplus \wedge^n E,$$

which just means that any $x \in \wedge E$ can be written in a unique way as a sum:

$$x = x_0 + x_1 + \cdots + x_n \ \text{ with } x_k \in \wedge^k E.$$

In fact, if we write $x = \sum_{J \in \hat{\mathcal{J}}} \lambda_J \boldsymbol{e}_{\hat{J}}$, then

$$x_k = \sum_{J \in \hat{\mathcal{J}}_k} \lambda_J \boldsymbol{e}_{\hat{J}}, \quad \text{where } \hat{\mathcal{J}}_k = \{J \in \hat{\mathcal{J}} : |J| = k\}.$$

Since $x_k \in \wedge^k E$ and $x_{k'} \in \wedge^{k'} E$ imply that $x_k \wedge x_{k'} \in \wedge^{k+k'} E$, $\wedge E$ is a *graded algebra*. Even more, the exterior product is *skew-commutative* (or *supercommutative*):

$$x_k \wedge x_{k'} = (-1)^{kk'} x_{k'} \wedge x_k. \tag{2.3}$$

The elements of $\wedge E$ are called *multivectors*, and those of $\wedge^k E$, *k-vectors*. The non-zero k-vectors of the form $\boldsymbol{x}_1 \wedge \cdots \wedge \boldsymbol{x}_k$ are called *k-blades* (alternatively, especially in mathematics texts, they are also called *decomposable* k-vectors). The 1-vectors are just the vectors, and 2-vectors and 3-vectors are usually called *bivectors* and *trivectors*, respectively. The elements of $\wedge^n E$ are called *pseudoscalars*. For any basis $\mathbf{e} = \boldsymbol{e}_1, \ldots, \boldsymbol{e}_n$ of E, we will say that $I_{\mathbf{e}} = \boldsymbol{e}_1 \wedge \cdots \wedge \boldsymbol{e}_n$ is the *pseudoscalar associated to* $\mathbf{e}$. Note that if x is a pseudoscalar, then $\boldsymbol{e} \wedge x = 0$ for any vector $\boldsymbol{e}$. Conversely, if $\boldsymbol{e} \wedge x = 0$ for any vector $\boldsymbol{e}$, then x is a pseudoscalar (see E.2.6, p. 38).

2.1.9 (Universal Property of $\wedge^k E$) If $f : E^k \to E'$ is a skew-symmetric multilinear map, where E' is any vector space, there is a unique linear map $\bar{f} : \wedge^k E \to E'$ such that $\bar{f}(\boldsymbol{x}_1 \wedge \cdots \wedge \boldsymbol{x}_k) = f(\boldsymbol{x}_1, \ldots, \boldsymbol{x}_k)$ for all $\boldsymbol{x}_1, \ldots, \boldsymbol{x}_k \in E$.

Proof Since f is multilinear, it is determined by the values of $f(\boldsymbol{e}_{j_1}, \ldots, \boldsymbol{e}_{j_k})$, for $1 \leqslant j_1, \ldots, j_k \leqslant n$. If f is skew-symmetric, then f is determined by the values $f(\boldsymbol{e}_{j_1}, \ldots, \boldsymbol{e}_{j_k})$, for $1 \leqslant j_1 < \cdots < j_k \leqslant n$, and the unique linear map $\bar{f} : \wedge^k E \to F$ such that $\bar{f}(\boldsymbol{e}_{j_1} \wedge \cdots \wedge \boldsymbol{e}_{j_k}) = f(\boldsymbol{e}_{j_1}, \ldots, \boldsymbol{e}_{j_k})$ $(1 \leqslant j_1 < \cdots < j_k \leqslant n)$ satisfies the claim. □

2.1.10 (Functoriality of $\wedge^k E$) With the notations of 2.1.5, the algebra homomorphism $f^{\wedge} : \wedge E \to \wedge E'$ extending the linear map $f : E \to E'$ induces a linear map $f^{\wedge k} : \wedge^k E \to \wedge^k E'$ that is determined by Eq. (2.2). If f is an endomorphism of E, then we have endomorphisms $f^{\wedge k} : \wedge^k E \to \wedge^k E$. In the case that $n = \dim E$, the space $\wedge^n E$ is one-dimensional and therefore there is a scalar δ such that $f^{\wedge n}(x) = \delta x$ for any $x \in \wedge^n E$. We say that δ is the *determinant* of f, and we write $\det(f)$ to denote it. This definition implies immediately the $\det(\mathrm{Id}_E) = 1$ and $\det(gf) = \det(g)\det(f)$ for any $f, g \in \mathrm{End}(E)$. If A is the matrix of f with respect to a basis $\mathbf{e} = \boldsymbol{e}_1, \ldots, \boldsymbol{e}_n$, then it is an straightforward exercise (see E.2.5, p. 38) to check that $\det(f) = \det(A)$. □

2.1.11 (Parity Involution of $\wedge E$) There exists a unique algebra *automorphism* of $\wedge E$, $x \mapsto \hat{x}$, such that $\hat{\boldsymbol{e}} = -\boldsymbol{e}$ for all $\boldsymbol{e} \in E$. For $x \in \wedge^k E$, we have $\hat{x} = (-1)^k x$. This automorphism is involutive, that is, $\hat{\hat{x}} = x$ for all x, and it is called the *parity involution* of $\wedge E$. □

2.1.12 (Reverse Involution of $\wedge E$) *There exists a unique algebra* anti-automorphism *of* $\wedge E$, $x \mapsto \tilde{x}$, *such that* $(\boldsymbol{x}_1 \wedge \cdots \wedge \boldsymbol{x}_k)^{\sim} = \boldsymbol{x}_k \wedge \cdots \wedge \boldsymbol{x}_1$ *for all* $k \geqslant 1$ *and* $\boldsymbol{x}_1, \ldots, \boldsymbol{x}_k \in E$. *It is clearly involutive, and it is called the* reverse involution *of* $\wedge E$.

Proof It is enough to apply 2.1.5 to the inclusion map $E \to \wedge^{\mathrm{op}} E$, where $\wedge^{\mathrm{op}} E$ is the opposite algebra of $\wedge E$ (this is the same algebra as $\wedge E$, but with the product $x \wedge^{\mathrm{op}} y$ defined as $y \wedge x$). □

2.1.13 (Remarks and Notations) We have used the same symbols for the parity and reverse involutions of $\wedge E$ than for the parity and reverse involutions of the tensor algebra TE. This corresponds to the fact that the involutions of TE leave invariant the ideal $\mathcal{I}$ of TE generated by the elements of the form $\boldsymbol{x} \otimes \boldsymbol{x}$, and hence they induce involutions of the quotient $\wedge E = TE/\mathcal{I}$ that coincide with the involutions of $\wedge E$ described before.

Since $\boldsymbol{x}_k \wedge \cdots \wedge \boldsymbol{x}_1 = (-1)^{\binom{k}{2}} \boldsymbol{x}_1 \wedge \cdots \wedge \boldsymbol{x}_k$, we have that $\tilde{x} = (-1)^{\binom{k}{2}} x = (-1)^{k/\!\!/2}$ for $x \in \wedge^k E$, where $k/\!\!/2$ denotes the integer quotient of k by 2, or $\lfloor k/2 \rfloor$. For this, we use that $\binom{k}{2}$ has the same parity as $k/\!\!/2$. Note that the sign $(-1)^{k/\!\!/2}$ is 1 (-1) if the *remainder* of k divided by 4 is 0 or 1 (2 or 3):

$$(4m)/\!\!/2 = (4m+1)/\!\!/2 = 2m \quad \text{and} \quad (4m+2)/\!\!/2 = (4m+3)/\!\!/2 = 2m+1.$$

2.1.14 (Blades and Linear Subspaces) The geometric relevance of the notion of k-blade stems from the following equivalence:

$$\boldsymbol{x} \in \langle \boldsymbol{x}_1, \ldots, \boldsymbol{x}_k \rangle \Leftrightarrow \boldsymbol{x} \wedge X = 0, \tag{2.4}$$

as it shows that X determines the linear subspace $\langle \boldsymbol{x}_1, \ldots, \boldsymbol{x}_k \rangle$. The converse is also true up to a non-zero scalar factor: *If $X' = \boldsymbol{x}'_1 \wedge \cdots \wedge \boldsymbol{x}'_k$ is another k-blade, then $\langle \boldsymbol{x}_1, \ldots, \boldsymbol{x}_k \rangle = \langle \boldsymbol{x}'_1, \ldots, \boldsymbol{x}'_k \rangle$ if and only if $X' \sim X$.*

Proof The if part follows immediately from Eq. (2.4). The only if part is a consequence of the fact that if $F = \langle \boldsymbol{x}_1, \ldots, \boldsymbol{x}_k \rangle = \langle \boldsymbol{x}'_1, \ldots, \boldsymbol{x}'_k \rangle$, then $X, X' \in \wedge^k F$ and hence $X' \sim X$ because $\wedge^k F$ is one-dimensional. □

2.1.15 (Remark) We see that $|X\rangle = |X'\rangle$ in the projective space $\mathbf{P}(\wedge^k E)$ of $\wedge^k E$ is equivalent to the equality of the corresponding subspaces. Henceforth, we will identify $|X\rangle$ with the linear subspace $\langle \boldsymbol{x}_1, \ldots, \boldsymbol{x}_k \rangle$ determined by X, so that $\boldsymbol{x} \in |X\rangle \Leftrightarrow \boldsymbol{x} \wedge X = 0$. In sum, the set $\mathrm{Gr}_k(E)$ of k-dimensional linear subspaces of E is naturally identified with a subset of the projective space $\mathbf{P}(\wedge^k E)$.

2.2 Contraction Operators for $\wedge E$

There is an analogue for the exterior powers $\wedge^k E$ and the exterior algebra $\wedge E$ of the contraction operators for the tensor algebra TE (see 1.4.17).

2.2.1 Given a linear form $\boldsymbol{\xi} \in E^*$, define, for any index $k \geqslant 1$, the map $E^k \to \wedge^{k-1} E$ by:

$$(\boldsymbol{y}_1, \ldots, \boldsymbol{y}_k) \mapsto \textstyle\sum_{j=1}^k (-1)^{j-1} \boldsymbol{\xi}(\boldsymbol{y}_j)\, \boldsymbol{y}_1 \wedge \cdots \wedge \boldsymbol{y}_{j-1} \wedge \boldsymbol{y}_{j+1} \wedge \cdots \otimes \boldsymbol{y}_k.$$

This map, which is clearly multilinear, is in fact skew-symmetric due to the signs $(-1)^{j-1}$. To see why, it may suffice to consider the cases $k = 2$ and $k = 3$:

$$i_\xi(\boldsymbol{y}_1 \wedge \boldsymbol{y}_2) = \xi(\boldsymbol{y}_1)\,\boldsymbol{y}_2 - \xi(\boldsymbol{y}_2)\,\boldsymbol{y}_1,$$
$$i_\xi(\boldsymbol{y}_1 \wedge \boldsymbol{y}_2 \wedge \boldsymbol{y}_3) = \xi(\boldsymbol{y}_1)\,\boldsymbol{y}_2 \wedge \boldsymbol{y}_3 - \xi(\boldsymbol{y}_2)\,\boldsymbol{y}_1 \wedge \boldsymbol{y}_3 + \xi(\boldsymbol{y}_3)\,\boldsymbol{y}_1 \wedge \boldsymbol{y}_2.$$

The first expression changes sign if $\boldsymbol{y}_1$ and $\boldsymbol{y}_2$ are swapped. The same happens with the second expression, for the sum of the first two summands and the third summand on the right-hand side are both skew-symmetric in $\boldsymbol{y}_1$ and $\boldsymbol{y}_2$. And, a similar reasoning holds when we swap $\boldsymbol{y}_2$ and $\boldsymbol{y}_3$. For the details in general, see E.2.7, p. 38.

Now by the universal property of $\wedge^k E$, there exists a unique linear map (there is no harm in using the same symbol as for the tensor algebra) $i_\xi : \wedge^k E \to \wedge^{k-1} E$ such that

$$i_\xi(\boldsymbol{y}_1 \wedge \cdots \wedge \boldsymbol{y}_k) = \textstyle\sum_{j=1}^k (-1)^{j-1} \xi(\boldsymbol{y}_j)\, \boldsymbol{y}_1 \wedge \cdots \wedge \boldsymbol{y}_{j-1} \wedge \boldsymbol{y}_{j+1} \wedge \cdots \wedge \boldsymbol{y}_k. \quad (2.5)$$

We also set, tipped by $\wedge^0 E = \mathbb{R}$ and $\wedge^{-1} E = \{0\}$, $i_\xi(\lambda) = 0$ for all $\lambda \in \mathbb{R}$. Thus, we get a linear map $i_\xi : \wedge E \to \wedge E$ of grade -1. This map, which is called the *ξ-contraction operator*, is a *skew-derivation*, which in this case means that

$$i_\xi(x \wedge y) = (i_\xi x) \wedge y + \hat{x} \wedge (i_\xi y) \quad (2.6)$$

for all $x, y \in \wedge E$. To prove this, it is enough to check the case in which x and y are wedge products of vectors, which in turn is an immediate consequence of Eq. (2.5).

If E is endowed with a metric q, then for any $\boldsymbol{x} \in E$ we define $i_{\boldsymbol{x}} = i_{q_{\boldsymbol{x}}}$ (the $\boldsymbol{x}$-contraction operator; recall that $q_{\boldsymbol{x}}(\boldsymbol{y}) = q(\boldsymbol{x}, \boldsymbol{y})$). It is thus clear that $i_{\boldsymbol{x}}$ is a grade -1 skew-derivation of $\wedge E$ which satisfies

$$i_{\boldsymbol{x}}(\boldsymbol{y}_1 \wedge \cdots \wedge \boldsymbol{y}_k) = \textstyle\sum_{j=1}^k (-1)^{j-1} q(\boldsymbol{x}, \boldsymbol{y}_j)\, \boldsymbol{y}_1 \wedge \cdots \wedge \boldsymbol{y}_{j-1} \wedge \boldsymbol{y}_{j+1} \wedge \cdots \wedge \boldsymbol{y}_k. \quad (2.7)$$

As for the tensor algebra contraction operators, it is an easy exercise to see that i_ξ ($i_{\boldsymbol{x}}$) is the unique skew-derivation of $\wedge E$ such that $i_\xi(\boldsymbol{y}) = \xi(\boldsymbol{y})$ ($i_\xi(\boldsymbol{y}) = q(\boldsymbol{x}, \boldsymbol{y})$) for any vector $\boldsymbol{y}$.

2.2.2 (Remarks) It is easily checked that the contraction operators i_ξ ($i_{\boldsymbol{x}}$) depend linearly on ξ (on $\boldsymbol{x}$). For example, if $\xi, \xi' \in E^*$ and $\lambda, \lambda' \in \mathbb{R}$, then

$$i_{\lambda\xi + \lambda'\xi'} = \lambda i_\xi + \lambda' i_{\xi'}.$$

On the other hand, $i_\xi^2 = 0$ for any linear form ξ (hence also $i_{\boldsymbol{x}}^2 = 0$ for any vector $\boldsymbol{x}$). Indeed, it is clear that i_ξ^2 vanishes on vectors, and it is easy to check that i_ξ^2 is a graded *derivation* of grade -2 (the essential point is that $\widehat{i_\xi y} = -i_\xi \hat{y}$ ($y \in \wedge E$), which holds because i_ξ decreases grades by 1; cf. E.2.8, p. 38), so i_ξ^2 vanishes on blades and consequently it is 0.

In particular, we can look at i_x as a linear endomorphism of $\wedge E$, and hence we get a linear map $E \to \mathrm{End}(\wedge E)$, $\boldsymbol{x} \mapsto i_{\boldsymbol{x}}$. Since $i_{\boldsymbol{x}}^2 = 0$ in the algebra $\mathrm{End}(\wedge E)$, by the universal property of the exterior algebra (see 2.1.1 (3)) there is a unique algebra homomorphism $\wedge E \to \mathrm{End}(\wedge E)$, $x \mapsto i_x$, that agrees with $\boldsymbol{x} \mapsto i_{\boldsymbol{x}}$ for vectors. Since it is an algebra homomorphism, in particular we have

$$i_{\boldsymbol{x}_1\wedge\cdots\wedge\boldsymbol{x}_k} = i_{\boldsymbol{x}_1}\cdots i_{\boldsymbol{x}_k}, \tag{2.8}$$

where the product on the right is the composition product of $\mathrm{End}(\wedge E)$. Since the $i_{\boldsymbol{x}_j}$ are graded of grade -1, $i_{\boldsymbol{x}_1\wedge\cdots\wedge\boldsymbol{x}_k}$ is graded of grade $-k$. This implies, more generally, that if $x \in \wedge^k E$ and $y \in \wedge^l E$, then $i_x(y) \in \wedge^{l-k} E$. Remark that the fact that $x \mapsto i_x$ is an algebra homomorphism implies that we also have $i_\lambda = \lambda \mathrm{Id}_{\wedge E}$, or $i_\lambda(y) = \lambda y$ for all $y \in \wedge E$ and $\lambda \in \mathbb{R}$.

The important question of how to evaluate $i_{\boldsymbol{x}_1\wedge\cdots\wedge\boldsymbol{x}_k}(y)$ when y is itself a blade is answered in next block. For $\boldsymbol{x}_1, \ldots, \boldsymbol{x}_k, \boldsymbol{y}_1, \ldots, \boldsymbol{y}_l \in E$, we set $X = \boldsymbol{x}_1 \wedge \cdots \wedge \boldsymbol{x}_k$, $Y = \boldsymbol{y}_1 \wedge \cdots \wedge \boldsymbol{y}_l$, and

$$Q(\boldsymbol{x}_1, \ldots, \boldsymbol{x}_k; \boldsymbol{y}_1, \ldots, \boldsymbol{y}_l) = \begin{pmatrix} q(\boldsymbol{x}_1, \boldsymbol{y}_1) & \cdots & q(\boldsymbol{x}_1, \boldsymbol{y}_l) \\ \vdots & & \vdots \\ q(\boldsymbol{x}_k, \boldsymbol{y}_1) & \cdots & q(\boldsymbol{x}_k, \boldsymbol{y}_l) \end{pmatrix}.$$

Given a multiindex $L \subseteq \{1, \ldots, l\}$, we set $Y_L = \wedge_{j\in L}\, \boldsymbol{y}_j$. Similarly, Q_L will denote the matrix $Q(\boldsymbol{x}_1, \ldots, \boldsymbol{x}_k; \boldsymbol{y}_j : j \in L)$.

2.2.3 (Laplace Rule)

(1) *With the above notations, we have*

$$\begin{aligned} i_X(Y) &= \textstyle\sum_M (-1)^{t(M,M')} \det(Q_M(\boldsymbol{x}_k, \ldots, \boldsymbol{x}_1; \boldsymbol{y}_1, \ldots, \boldsymbol{y}_l))Y_{M'} \\ &= (-1)^{k/\!/2} \textstyle\sum_M (-1)^{t(M,M')} \det(Q_M(\boldsymbol{x}_1, \ldots, \boldsymbol{x}_k; \boldsymbol{y}_1, \ldots, \boldsymbol{y}_l))Y_{M'}, \end{aligned}$$

where $M \subseteq \{1, \ldots, l\}$ has grade k and $M' = \{1, \ldots, l\} - M$.

(2) *For $l = k$, we have $i_X(Y) = (-1)^{k/\!/2} \det Q(\boldsymbol{x}_1, \ldots, \boldsymbol{x}_k; \boldsymbol{y}_1, \ldots, \boldsymbol{y}_k) = i_Y(X)$ because the determinant in this expression and* $\det Q(\boldsymbol{y}_1, \ldots, \boldsymbol{y}_k; \boldsymbol{x}_1, \ldots, \boldsymbol{x}_k)$ *are transposed.*

(3) *Finally, let us also remark that $i_X(X) = (-1)^{k/\!/2} \det Q(\boldsymbol{x}_1, \ldots, \boldsymbol{x}_k)$, where*

$$Q(\boldsymbol{x}_1, \ldots, \boldsymbol{x}_k) = Q(\boldsymbol{x}_1, \ldots, \boldsymbol{x}_k; \boldsymbol{x}_1, \ldots, \boldsymbol{x}_k).$$

Proof We only need to prove the first expression in (1), because Q is skew-symmetric in $\boldsymbol{x}_1, \ldots, \boldsymbol{x}_k$ and so the reordering of $\boldsymbol{x}_k, \ldots, \boldsymbol{x}_1$ to $\boldsymbol{x}_1, \ldots, \boldsymbol{x}_k$ amounts to $\binom{k}{2}$ sign changes, which is the same as $k/\!/2$ sign changes.

We will proceed by induction on k. For $k = 1$, the formula agrees with Eq. (2.5) (notice that the Y_j in that formula here are denoted by $Y_{\{1,\ldots,l\}-\{j\}}$). If $k > 1$, the expression (2.8) yields that $i_X(Y) = i_{X'}(i_{\boldsymbol{x}_k}Y)$, where $X' = \boldsymbol{x}_1 \wedge \cdots \wedge \boldsymbol{x}_{k-1}$. Now, $i_{\boldsymbol{x}_k}(Y) = \sum_{j=1}^{l}(-1)^{j-1}q(\boldsymbol{x}_k, \boldsymbol{y}_j)Y_{j'}$, where for simplicity we write $j' = \{1,\ldots,l\} - \{j\}$, and hence $i_X(Y) = \sum_{j=1}^{l}(-1)^{j-1}q(\boldsymbol{x}_k, \boldsymbol{y}_j)\, i_{X'}(Y_{j'})$. Now by induction on k, we have, setting $Q' = Q(\boldsymbol{x}_{k-1},\ldots,\boldsymbol{x}_1; \boldsymbol{y}_1,\ldots,\boldsymbol{y}_l)$:

$$i_X(Y) = \textstyle\sum_{j=1}^{l}(-1)^{j-1}q(\boldsymbol{x}_k, \boldsymbol{y}_j)\sum_L(-1)^{t(L,j'-L)}\det(Q'_L)\, Y_{j'-L},$$

where L runs over the grade $k-1$ multiindices in j'. On noticing that such multiindices L are in one-to-one correspondence with the grade k multiindices M of $\{1,\ldots,l\}$ that contain j (through the relations $L = M - \{j\}$ and $M = \{j\} + L$, where $\{j\} + L$ denotes the result of reordering $\{j\} \cup L$ in increasing order), we can swap the two summations in the last displayed formula and obtain

$$i_X(Y) = \textstyle\sum_M\sum_{j\in M}(-1)^{j-1}(-1)^{t(L,M')}q(\boldsymbol{x}_k, \boldsymbol{y}_j)\det(Q'_L)Y_{M'}.$$

For a given $M = \{m_1,\ldots,m_k\} \subseteq \{1,\ldots,l\}$ and $j \in M$, let $i \in \{1,\ldots,k\}$ be the index such that $j = m_i$. Then, the sign in the last expression depends on the parity of

$$m_i - 1 + t(M - \{m_i\}, M') = m_i - 1 - t(m_i, M') + t(M, M') = i - 1 + t(M, M'),$$

for $t(m_i, M') = t(m_i, \{1,\ldots,l\}) - t(m_i, \{m_1,\ldots,m_{i-1}\}) = (m_i - 1) - (i - 1)$. Thus:

$$i_X(Y) = \textstyle\sum_M\left(\sum_{i=1}^{l}(-1)^{i-1}q(\boldsymbol{x}_k, \boldsymbol{y}_{m_i})\det(Q'_{M-\{m_i\}})\right)Y_{M'}.$$

Finally, note that the expression inside the big parentheses is the development of $\det Q(\boldsymbol{x}_k, \boldsymbol{x}_{k-1},\ldots,\boldsymbol{x}_1; \boldsymbol{y}_{m_1},\ldots,\boldsymbol{y}_{m_k}) = \det Q_M(\boldsymbol{x}_k, \boldsymbol{x}_{k-1},\ldots,\boldsymbol{x}_1; \boldsymbol{y}_1,\ldots,\boldsymbol{y}_l)$ along the first row. □

2.3 The Metric Grassmann Algebra

The metric q of E induces a metric of $\wedge E$, which we will denote by the same symbol q. This metric is determined by bilinearity and the following two rules for $x \in \wedge^k E$ and $y \in \wedge^l E$:

(1) $q(x, y) = 0$ if $k \neq l$, and
(2) $q(x, y) = i_{\tilde{x}}(y)$ if $l = k$.

This extends the metric of E, as $i_{\tilde{x}}(y) = i_x(y) = q(x, y)$, and defines indeed a metric of $\wedge E$: it is bilinear, because $i_{\tilde{x}}(y)$ is a bilinear expression of x and y, and it is symmetric as a consequence of 2.2.3 (2), because for k-blades (using the same notations as in that statement)

$$q(X, Y) = i_{\tilde{X}}(Y) = (-1)^{k/\!/2} i_X(Y) = \det Q(\boldsymbol{x}_1, \ldots, \boldsymbol{x}_k; \boldsymbol{y}_1, \ldots, \boldsymbol{y}_k) \tag{2.9}$$

and the claim follows because

$$Q(\boldsymbol{y}_1, \ldots, \boldsymbol{y}_k; \boldsymbol{x}_1, \ldots, \boldsymbol{x}_k) \text{ and } Q(\boldsymbol{x}_1, \ldots, \boldsymbol{x}_k; \boldsymbol{y}_1, \ldots, \boldsymbol{y}_k)$$

are transposed matrices. It is worth remarking that for $Y = X$ we get

$$q(X) = G(\boldsymbol{x}_1, \ldots, \boldsymbol{x}_k) := \begin{vmatrix} q(\boldsymbol{x}_1, \boldsymbol{x}_1) & \cdots & q(\boldsymbol{x}_1, \boldsymbol{x}_k) \\ \vdots & & \vdots \\ q(\boldsymbol{x}_k, \boldsymbol{x}_1) & \cdots & q(\boldsymbol{x}_k, \boldsymbol{x}_k) \end{vmatrix}, \tag{2.10}$$

where we write $G(\boldsymbol{x}_1, \ldots, \boldsymbol{x}_k) = \det Q(\boldsymbol{x}_1, \ldots, \boldsymbol{x}_k)$. We will refer to Eq. (2.10) as *Gram's formula*.

2.3.1 (Example: Orthogonal Blades and Orthogonal Basis of $\wedge E$) Equation (2.9) implies that if $\boldsymbol{x}_j$ is orthogonal to all $\boldsymbol{y}_1, \ldots, \boldsymbol{y}_k$, then $q(X, Y) = 0$. In particular, we see that if $\mathbf{e}$ is an orthogonal basis of E, then the basis $\{\boldsymbol{e}_{\hat{J}}\}$ of $\wedge E$ (see 2.1.6) is orthogonal. Moreover, by Eq. (2.10) we get

$$q(\boldsymbol{e}_{\hat{J}}) = q_J, \quad \text{where } q_J = q(\boldsymbol{e}_{j_1}) \cdots q(\boldsymbol{e}_{j_k}). \tag{2.11}$$

□

2.3.2 (Vanishing of $q(X)$) *For a non-zero k-blade $X = \boldsymbol{x}_1 \wedge \ldots \wedge \boldsymbol{x}_k$, $q(X) = 0$ if and only if the space $|X\rangle$ is singular.*

Proof For any basis $\mathbf{u} = \boldsymbol{u}_1, \ldots, \boldsymbol{u}_k$ of $|X\rangle$, $X = \delta U$, $U = \boldsymbol{u}_1 \wedge \ldots \wedge \boldsymbol{u}_k$ and δ a non-zero scalar. In particular, $q(X) = \delta^2 q(U)$. If X is regular, we can choose $\mathbf{u}$ orthonormal, so that $q(U) = \pm 1$, and hence $q(X) \neq 0$. If X is singular, then we can choose $\mathbf{u}$ so that $\boldsymbol{u}_1$ is orthogonal to all vectors in $|X\rangle$, in which case the first row of $Q(\boldsymbol{u}_1, \ldots, \boldsymbol{u}_k)$ is zero, so $q(U) = 0$, and hence $q(X) = 0$. □

2.3.3 *If (E, q) is Euclidean* (positive definite), *then*

$$\det Q(\boldsymbol{x}_1, \ldots, \boldsymbol{x}_k) = V(\boldsymbol{x}_1, \ldots, \boldsymbol{x}_k)^2,$$

where $V(\boldsymbol{x}_1, \ldots, \boldsymbol{x}_k)$ is the k-volume of the parallelepiped defined by $\boldsymbol{x}_1, \ldots, \boldsymbol{x}_k$. Thus, $\wedge E$ is also Euclidean.

Proof The formula is true for $k = 1$, as the 1-volume of $\boldsymbol{x}_1$ is $|\boldsymbol{x}_1|$ and $|\boldsymbol{x}_1|^2 = q(\boldsymbol{x}_1) = G(\boldsymbol{x}_1)$. For $k > 1$, we can first proceed by induction to show that the formula is true when $\boldsymbol{x}_k$ is orthogonal to $\langle \boldsymbol{x}_1, \ldots, \boldsymbol{x}_{k-1}\rangle$. Indeed, in this case:

$$G(\boldsymbol{x}_1, \ldots, \boldsymbol{x}_k) = G(\boldsymbol{x}_1, \ldots, \boldsymbol{x}_{k-1})q(\boldsymbol{x}_k) = V(\boldsymbol{x}_1, \ldots, \boldsymbol{x}_{k-1})^2|\boldsymbol{x}_k|^2$$
$$= V(\boldsymbol{x}_1, \ldots, \boldsymbol{x}_k)^2.$$

To show the validity of the formula in general, decompose $\boldsymbol{x}_k = \boldsymbol{x}'_k + \boldsymbol{x}''_k$ such that $\boldsymbol{x}'_k \in \langle \boldsymbol{x}_1, \ldots, \boldsymbol{x}_{k-1}\rangle$ and $\boldsymbol{x}''_k \in \langle \boldsymbol{x}_1, \ldots, \boldsymbol{x}_{k-1}\rangle^{\perp}$. With this, we easily see that

$$G(\boldsymbol{x}_1, \ldots, \boldsymbol{x}_k) = G(\boldsymbol{x}_1, \ldots, \boldsymbol{x}_{k-1})q(\boldsymbol{x}''_k) + G(\boldsymbol{x}_1, \ldots, \boldsymbol{x}_{k-1}, \boldsymbol{x}'_k),$$

and this reduces to $G(\boldsymbol{x}_1, \ldots, \boldsymbol{x}_{k-1})q(x''_k)$, because $\boldsymbol{x}'_k \in \langle \boldsymbol{x}_1, \ldots, \boldsymbol{x}_{k-1}\rangle$ and this implies that $G(\boldsymbol{x}_1, \ldots, \boldsymbol{x}_{k-1}, \boldsymbol{x}'_k) = 0$ (see E.2.9, p. 39). Now, $G(\boldsymbol{x}_1, \ldots, \boldsymbol{x}_{k-1}) = V(\boldsymbol{x}_1, \ldots, \boldsymbol{x}_{k-1})^2$ by induction, $q(\boldsymbol{x}''_k) = |\boldsymbol{x}''_k|^2$ is the n-th height of the parallelepiped $[\boldsymbol{x}_1, \ldots, \boldsymbol{x}_k]$ over the base $[\boldsymbol{x}_1, \ldots, \boldsymbol{x}_{k-1}]$, and hence

$$G(\boldsymbol{x}_1, \ldots, \boldsymbol{x}_{k-1})q(\boldsymbol{x}''_k) = V(\boldsymbol{x}_1, \ldots, \boldsymbol{x}_k)^2.$$

□

It is a curious fact that if $s > 0$, then the signature of $\wedge E_{r,s}$ is always $(2^{n-1}, 2^{n-1})$ (see E.2.10, p. 39).

2.3.4 (Pythagoras Theorem) If $\boldsymbol{x} = \lambda_1\boldsymbol{e}_1 + \cdots + \lambda_n\boldsymbol{e}_n$, where $\boldsymbol{e}_1, \ldots, \boldsymbol{e}_n$ is an orthonormal basis of E_n, then we have

$$\boldsymbol{x}^2 = \lambda_1^2 + \cdots + \lambda_n^2.$$

Since $\boldsymbol{x}_j = \lambda_j\boldsymbol{e}_j$ is the orthogonal projection of $\boldsymbol{x}$ to the j-th axis $\langle \boldsymbol{e}_j\rangle$, and $\boldsymbol{x}_j^2 = \lambda_j^2$, the displayed formula is entitled to be called *Pythagoras' theorem* for lengths of the Euclidean space E_n. Actually, there is a Pythagoras theorem for extensions x of any grade k. Indeed, we have an expansion $x = \sum_J \lambda_J\boldsymbol{e}_{\hat{J}}$, the $\boldsymbol{e}_{\hat{J}}$ form an orthonormal basis of $\wedge^k E_n$, $x_J = \lambda_J\boldsymbol{e}_{\hat{J}}$ is the orthogonal projection of x to the "axis" $\langle \boldsymbol{e}_{\hat{J}}\rangle$, and $q(x) = \sum_J \lambda_J^2 = \sum_J q(x_J)$.

2.4 The Inner Product

Let us begin with a remark on the *right contraction* of a vector $\boldsymbol{x}$ with a blade $Y = \boldsymbol{y}_1 \wedge \cdots \wedge \boldsymbol{y}_k$, which is defined as the left contraction, but by running over the $\boldsymbol{y}_j$ in reverse order and starting with the sign $+1$ for $\boldsymbol{y}_k$. What we get is

$$(-1)^{k+1}i_{\boldsymbol{x}}(Y).$$

Indeed, the right contraction is given by $\sum_{j=1}^{k}(-1)^{j-1}q(\boldsymbol{x}, \boldsymbol{y}_{k+1-j})Y_{k+1-j}$, which is the same (set $l = k+1-j$) as $\sum_{l=1}^{k}(-1)^{k-l}q(\boldsymbol{x}, \boldsymbol{y}_l)Y_l = (-1)^{k+1}i_{\boldsymbol{x}}(Y)$.

The *inner product* of $\wedge E$, which is denoted by $x \cdot y$, is defined for homogeneous multivectors and then extended bilinearly to all multivectors. For $x \in \wedge^k E$, $y \in \wedge^l E$, the definition is as follows:

$$x \cdot y = \begin{cases} 0 & \text{if } k = 0 \text{ or } l = 0, \\ i_x(y) & \text{if } 0 < k \leqslant l, \\ (-1)^{kl+l} i_y(x) & \text{if } 0 < l \leqslant k. \end{cases} \quad (2.12)$$

As we will see, the role of the first rule is to make sure that other important formulas that we will establish later on do not have exceptions. The third rule is like the second, but using right contractions rather than left contractions. For $k = l$, both rules give the same answer by 2.2.3 (2). Beware, however, that the inner product is not associative, nor unital, and in general it is not commutative.

2.4.1 (Recursive Formulas) *If $k, l \geqslant 2$ and $\boldsymbol{x}_1, \ldots, \boldsymbol{x}_k, \boldsymbol{y}_1, \ldots, \boldsymbol{y}_l \in E$, then*

$$(\boldsymbol{x}_1 \wedge \cdots \wedge \boldsymbol{x}_k) \cdot (\boldsymbol{y}_1 \wedge \cdots \wedge \boldsymbol{y}_l) = \begin{cases} (\boldsymbol{x}_1 \wedge \cdots \wedge \boldsymbol{x}_{k-1}) \cdot (\boldsymbol{x}_k \cdot Y) & \text{if } k \leqslant l, \\ (X \cdot \boldsymbol{y}_1) \cdot (\boldsymbol{y}_2 \wedge \cdots \wedge \boldsymbol{y}_l) & \text{if } k \geqslant l, \end{cases} \quad (2.13)$$

where we set $X = \boldsymbol{x}_1 \wedge \cdots \wedge \boldsymbol{x}_k$ and $Y = \boldsymbol{y}_1 \wedge \cdots \wedge \boldsymbol{y}_l$.

Proof The first rule is an immediate consequence of Eq. (2.8) and the second follows from the first and the commutation relations. In detail,

$$\begin{aligned} &(\boldsymbol{x}_1 \wedge \cdots \wedge \boldsymbol{x}_k) \cdot (\boldsymbol{y}_1 \wedge \cdots \wedge \boldsymbol{y}_l) \\ &= (-1)^{(k+1)l} (\boldsymbol{y}_1 \wedge \cdots \wedge \boldsymbol{y}_l) \cdot (\boldsymbol{x}_1 \wedge \cdots \wedge \boldsymbol{x}_k) \\ &= (-1)^{(k+1)l+l-1} (\boldsymbol{y}_2 \wedge \cdots \wedge \boldsymbol{y}_l \wedge \boldsymbol{y}_1) \cdot (\boldsymbol{x}_1 \wedge \cdots \wedge \boldsymbol{x}_k) \\ &= (-1)^{(k+1)l+l-1} (\boldsymbol{y}_2 \wedge \cdots \wedge \boldsymbol{y}_l) \cdot (\boldsymbol{y}_1 \cdot (\boldsymbol{x}_1 \wedge \cdots \wedge \boldsymbol{x}_k)) \\ &= (-1)^{(k+1)l+l-1+k+1} (\boldsymbol{y}_2 \wedge \cdots \wedge \boldsymbol{y}_l) \cdot ((\boldsymbol{x}_1 \wedge \cdots \wedge \boldsymbol{x}_k) \cdot \boldsymbol{y}_1) \\ &= ((\boldsymbol{x}_1 \wedge \cdots \wedge \boldsymbol{x}_k) \cdot \boldsymbol{y}_1) \cdot (\boldsymbol{y}_2 \wedge \cdots \wedge \boldsymbol{y}_l), \end{aligned}$$

where in the last step we use that $(k+1)l + l - 1 + k + 1 \equiv k(l-1) \bmod 2$. □

2.4.2 (Remark on the Commutation Rule of the Inner Product) The third expression of Eq. (2.12) tells us that $x \cdot y = (-1)^{kl+l} y \cdot x$ when $l \leqslant k$. This rule is valid in general under the form $x \cdot y = (-1)^{kl+m} y \cdot x$, where $m = \min(k, l)$. Indeed, if $k \leqslant l$, and hence $m = k$, then $y \cdot x = (-1)^{kl+k} x \cdot y$ (by the third case in Eq. (2.12)) and consequently $x \cdot y = (-1)^{kl+m} y \cdot x$.

In particular, we see that the inner product is symmetric when $k = l$. In general, it is symmetric if k and l have the same parity or else when m is even. Otherwise (namely when k and l have different parity and m is odd), it is skew-symmetric.

2.4.3 (Remark on the First Rule (2.12)**)** This rule stipulates that $\lambda \cdot x = x \cdot \lambda = 0$. So in the second rule (2.12), the condition $k > 0$ is crucial, as for $k = 0$ it would give $\lambda \cdot y = i_\lambda(y) = \lambda y$, as noticed before (just below Eq. (2.8)). Similarly, in the third rule in Eq. (2.12) the condition $l > 0$ is required, because for $l = 0$ the rule would give $x \cdot \lambda = i_\lambda(x) = \lambda x$.

2.4.4 (The Parity Involution Is an Automorphism of the Inner Product) *If $x, y \in \wedge E$, then $\widehat{x \cdot y} = \hat{x} \cdot \hat{y}$.*

Proof Bilinearity tells us that it is enough to establish the relation for $x \in \wedge^k E$ and $y \in \wedge^l E$. Since the inner product vanishes if $k = 0$ or $l = 0$, we can also assume that $k, l > 0$. Now, the left-hand side of the claimed relation is $(-1)^{|k-l|} x \cdot y$, and the right-hand side is $(-1)^k(-1)^l x \cdot y$, so the conclusion follows from the fact that $|k - l|$ and $k + l$ have the same parity. □

2.4.5 (The Reverse of an Inner Product) *If $x, y \in \wedge E$, then $\widetilde{x \cdot y} = \tilde{y} \cdot \tilde{x}$.*

Proof We can work under the same assumptions as in the previous proof. Let $m' = \max(k, l)$ and $m = \min(k, l)$. Then:

$$\widetilde{x \cdot y} = (-1)^{(m'-m)/\!/2} x \cdot y,$$

whereas

$$\tilde{y} \cdot \tilde{x} = (-1)^{l/\!/2}(-1)^{k/\!/2} y \cdot x = (-1)^{l/\!/2}(-1)^{k/\!/2}(-1)^{kl+m} x \cdot y.$$

So all boils down to see that $(m' - m)/\!/2$ and $l/\!/2 + k/\!/2 + kl + m$ have the same parity, which can be checked in a straightforward way (cf. E.2.11, p. 39). □

2.4.6 (Laplace's Rule in Terms of the Inner Product) In the case of two blades, say $X \in \wedge^k E$ and $Y = \boldsymbol{y}_1 \wedge \cdots \wedge \boldsymbol{y}_l \in \wedge^l E$, the definition of the inner product in the case $k \leqslant l$ (Eq. (2.12)) and the definition of the metric in the case $k = l$ (Eq. (2.9)) allow us to reinterpret the Laplace rule 2.2.3 in the following way:

$$X \cdot Y = \textstyle\sum_M (-1)^{t(M,M')} q(\tilde{X}, Y_M) Y_{M'} = \textstyle\sum_M (-1)^{t(M,M')} (X \cdot Y_M) Y_{M'}, \qquad (2.14)$$

where $M = m_1, \ldots, m_k$ runs through all weight k multiindices in $\{1, \ldots, l\}$,

$$M' = \{1, \ldots, l\} - M, \quad Y_M = \boldsymbol{y}_{m_1} \wedge \cdots \wedge \boldsymbol{y}_{m_k},$$

and with $Y_{M'}$ formed in a similar way with M'. For a few concrete examples of Laplace's rule, see E.2.12, p. 39 □

An immediate consequence is that in the case where $k = l$, the inner product can be expressed in terms of the metric, and vice versa:

2.4.7 (Metric Formulas) *If $x, y \in \wedge^k E$, $x \cdot y = q(\tilde{x}, y)$, $q(x, y) = \tilde{x} \cdot y$.* □

2.4.8 (Metric Norm) *If* $x \in \wedge^k E$, $q(x) = \tilde{x} \cdot x = (-1)^{k/\!/2} x \cdot x$. □

2.4.9 (Inner Product Table) Another important consequence of Eq. (2.14) is the expression of the inner products $e_{\hat{K}} \cdot e_{\hat{L}}$ for the basis $\hat{B} = \{e_{\hat{J}}\}_{J \subseteq N}$ of $\wedge E$ associated to an orthogonal basis $\mathbf{e} = e_1, \ldots, e_n$ of E. To determine this product, the commutation rule for the inner product allows us to assume that $k \leqslant l$, where $k = |K|$ and $l = |L|$.

Since $\hat{B}$ is q-orthogonal, Eq. (2.14) tells us that $e_{\hat{K}} \cdot e_{\hat{L}} = 0$ unless $K \subseteq L$, and in this case we have

$$e_{\hat{K}} \cdot e_{\hat{L}} = (-1)^{k/\!/2}(-1)^{t(K,L-K)} q(e_{\hat{K}}) e_{(L-K)\hat{}} = (-1)^{t(K,L)} q(e_{\hat{K}}) e_{(L-K)\hat{}}.$$

In the case $k \geqslant l$, using the commutation rule of the inner product we get a similar result, namely $e_{\hat{K}} \cdot e_{\hat{L}} = (-1)^{t(L,K)} q(e_{\hat{L}}) e_{(K-L)\hat{}}$ if $L \subseteq K$ and 0 otherwise (see E.2.13, p. 40).To sum up, we have

$$e_{\hat{K}} \cdot e_{\hat{L}} = \begin{cases} (-1)^{t(K,L)} q(e_{\hat{K}}) e_{(L-K)\hat{}} & \text{if } K \subseteq L, \\ (-1)^{t(K,L)} q(e_{\hat{L}}) e_{(K-L)\hat{}} & \text{if } L \subseteq K, \\ 0 & \text{otherwise.} \end{cases} \tag{2.15}$$

We end the section with three results that will be useful later on.

2.4.10 (Vanishing of the Inner Product) *Let X be a k-blade and Y an l-blade. Then, $X \cdot Y = 0$ if $k \leqslant l$, and there is a non-zero vector in $|X\rangle$ that is orthogonal to $|Y\rangle$ or if $k \geqslant l$ and there is a non-zero vector in $|Y\rangle$ that is orthogonal to $|X\rangle$.*

Proof By the commutation relation of the inner product of blades, it is enough to prove the first claim. Suppose, then, that $k \leqslant l$, and that there is non-zero vector in $|X\rangle$ that is orthogonal to $|Y\rangle$. These hypotheses imply that we can find a basis $v_1, \ldots, v_k$ of $|X\rangle$ such that v_k is orthogonal to $|Y\rangle$. So, we have $v_k \cdot Y = 0$ and hence also, by the recursive relations, $V \cdot Y = 0$, where $V = v_1 \wedge \cdots \wedge v_k$. Now, the relation $X \cdot Y = 0$ follows because $X \sim V$. □

2.4.11 *Let $x \in \wedge E$. If $e \cdot x = 0$ for all $e \in E$, then x is a scalar.*

Proof Let $e_1, \ldots, e_n \in E$ be an orthogonal basis and write $x = \sum_J \lambda_J e_{\hat{J}}$, where J runs over all multiindices from $\{1, \ldots, n\}$. Our hypothesis implies $e_1 \cdot x = 0$. But, $e_1 \cdot e_{\hat{J}} = q(e_1) e_{\widehat{J'}}$, $J' = J - \{1\}$, if $1 \in J$, and $e_1 \cdot e_{\hat{J}} = 0$ if $1 \notin J$. Thus, we get $0 = \sum_{J, 1 \in J} \lambda_J e_{\widehat{J'}}$, and hence $\lambda_J = 0$ for all J such that $1 \in J$. This just says that $x \in \wedge\langle e_2, \ldots, e_n\rangle$. Now, we can proceed in the same way with $e_2, \ldots, e_n$ and conclude successively that no e_k ($k = 1, \ldots, n$) appears in the expansion of x, so $x = x_0 \in \mathbb{R}$. □

Note the analogy of this result with E.2.6, p. 38.

2.5 Exercises

E.2.1 With the notations of Example 2.1.4, show that

$$(\delta_I \wedge \delta_J) \wedge \delta_K = \delta_I \wedge (\delta_J \wedge \delta_K) \text{ for all } I, J, K \in \hat{\mathcal{J}},$$

and that this implies that the exterior product of $\mathbb{R}[\hat{\mathcal{J}}]$ is associative.

Hint Show that both expressions are 0 unless $I \cap J = I \cap K = J \cap K = \emptyset$, and that if this is the case, then both sides are equal to $(-1)^{t(I,J,K)}\delta_L$, where L is the result of arranging $I \cup J \cup K$ in increasing order.

E.2.2 (The Algebra $\mathbb{R}[\hat{\mathcal{J}}]$) With the notations of Example 2.1.4, we want to show that $(\mathbb{R}[\hat{\mathcal{J}}], \wedge)$ satisfies the axioms 2.1.1 for $E' = \langle \delta_1, \ldots, \delta_n \rangle$. Since the properties (1) and (2) are clear (recall Eq. (2.1)), it only remains to prove (3).

Let $f : E' \to \mathcal{A}$ be a linear map from E' into an algebra $(\mathcal{A}, *)$ and suppose that $(f\boldsymbol{x}')^{*2} = 0$ for all $\boldsymbol{x}' \in E'$. If for any $J \in \hat{\mathcal{J}}$ we set $e'_J = f(\delta_{j_1}) * \cdots * f(\delta_{j_k})$, then there is unique linear map $f^* : \mathbb{R}[\hat{\mathcal{J}}] \to \mathcal{A}$ such that $\delta_J \mapsto e'_J$ (use the principle 1.2.4). To see that this map is an algebra homomorphism, the key point is to establish that $f^*\delta_J = f(\delta_{j_1}) * \cdots * f(\delta_{j_k})$ *for any index sequence* J. Indeed, the left-hand side is equal to $f^*\left((-1)^{t(J)}\delta_{\tilde{J}}\right) = (-1)^{t(J)}e'_{\tilde{J}}$, and this agrees with the right-hand side because any two elements among $f(\delta_{j_1}), \ldots, f(\delta_{j_k})$ anticommute. Now, we have, for $J, J' \in \hat{\mathcal{J}}$,

$$\begin{aligned} f^*(\delta_J \wedge \delta_{J'}) &= f^*(\delta_{J,J'}) = f(\delta_{j_1}) * \cdots * f(\delta_{j_k}) * f(\delta_{j'_1}) * \cdots * f(\delta_{j'_{k'}}) \\ &= f^*(\delta_J) * f^*(\delta_{J'}) \end{aligned}$$

and this implies that f^* is an algebra homomorphism extending f.

E.2.3 (Uniqueness of the Exterior Algebra) If $(\wedge E, \wedge)$ and $(\wedge' E, \wedge')$ satisfy the conditions of 2.1.1, then the inclusion of E in $\wedge' E$ satisfies $\boldsymbol{x}^{\wedge' 2} = \boldsymbol{x} \wedge' \boldsymbol{x} = 0$, so it can be extended uniquely to an algebra homomorphism $\wedge E \to \wedge' E$, and this homomorphism satisfies

$$\boldsymbol{x}_1 \wedge \cdots \wedge \boldsymbol{x}_k \mapsto \boldsymbol{x}_1 \wedge' \cdots \wedge' \boldsymbol{x}_k$$

for any positive integer k and any $\boldsymbol{x}_1, \ldots, \boldsymbol{x}_k \in E$. *This homomorphism is an isomorphism*, for there is also, for the same reasons, an algebra homomorphism $\wedge' E \to \wedge E$ such that $\boldsymbol{x}_1 \wedge' \cdots \wedge' \boldsymbol{x}_k \mapsto \boldsymbol{x}_1 \wedge \cdots \wedge \boldsymbol{x}_k$, and then the compositions $\wedge E \to \wedge' E \to \wedge E$ and $\wedge' E \to \wedge E \to \wedge' E$ are the identity because they satisfy $\boldsymbol{x}_1 \wedge \cdots \wedge \boldsymbol{x}_k \mapsto \boldsymbol{x}_1 \wedge \cdots \wedge \boldsymbol{x}_k$ and $\boldsymbol{x}_1 \wedge' \cdots \wedge' \boldsymbol{x}_k \mapsto \boldsymbol{x}_1 \wedge' \cdots \wedge' \boldsymbol{x}_k$, respectively.

E.2.4 (Existence of the Exterior Algebra) With the notations of 2.1.1, consider the map $i : E \to \wedge E$ defined as the composition of the linear inclusion $E \to TE$ followed by the quotient map $\pi : TE \to \wedge E = TE/\mathcal{I}$. Let $\bar{E} = i(E)$. Show that: (1) The map i is one-to-one and hence that $i : E \to \bar{E}$ is a linear isomorphism. This allows us to identify E and $\bar{E}$, as we will do in what follows. So, E is a linear subspace of $\wedge E$, and we have $\boldsymbol{x} \wedge \boldsymbol{x} = \pi(\boldsymbol{x} \otimes \boldsymbol{x}) = 0$ for all $\boldsymbol{x} \in E$. (2) The form of the generators of $\mathcal{I}$ implies that $1 \notin E$. Moreover, since E generates TE, $\wedge E$ is also generated by E. (3) As already shown in 2.1.8, $(\wedge E, \wedge)$ satisfies the third axiom 2.1.1.

E.2.5 ($\det(f) = \det(A)$) If $\mathbf{e} = \boldsymbol{e}_1, \ldots, \boldsymbol{e}_n$ is a basis of E, then on one hand we have $f^{\wedge n}(\boldsymbol{e}_1 \wedge \cdots \wedge \boldsymbol{e}_n) = \det(f)\, \boldsymbol{e}_1 \wedge \cdots \wedge \boldsymbol{e}_n$, by the definition of $\det(f)$, but on the other we have that $f^{\wedge n}(\boldsymbol{e}_1 \wedge \cdots \wedge \boldsymbol{e}_n) = (f\boldsymbol{e}_1) \wedge \cdots \wedge (f\boldsymbol{e}_n)$. Show that this expression is equal to $\det(A)\, \boldsymbol{e}_1 \wedge \ldots \wedge \boldsymbol{e}_n$, where $A = (a_j^k)$ is the matrix of f with respect to $\mathbf{e}$.

Hint Replace $f\boldsymbol{e}_j$ by $\sum_k a_j^k \boldsymbol{e}_k$ and expand $(f\boldsymbol{e}_1) \wedge \cdots \wedge (f\boldsymbol{e}_n)$ using the properties of the exterior product. For example, in the case $n = 2$ we have: $(a_1^1 \boldsymbol{e}_1 + a_1^2 \boldsymbol{e}_2) \wedge (a_2^1 \boldsymbol{e}_1 + a_2^2 \boldsymbol{e}_2) = a_1^1 a_2^2 \boldsymbol{e}_1 \wedge \boldsymbol{e}_2 + a_1^2 a_2^1 \boldsymbol{e}_2 \wedge \boldsymbol{e}_1 = (a_1^1 a_2^2 - a_1^2 a_2^1)\boldsymbol{e}_1 \wedge \boldsymbol{e}_2$.

E.2.6 Show that if a multivector x satisfies $\boldsymbol{e} \wedge x = 0$ for all vectors $\boldsymbol{e}$, then x is a pseudoscalar.

Hint If $x = \sum_J \lambda_J \boldsymbol{e}_{\hat{J}}$ $(*)$ is the expression of x in terms of the basis $\{\boldsymbol{e}_{\hat{J}}\}$ of $\wedge E$ associated to a basis $\boldsymbol{e}_1, \ldots, \boldsymbol{e}_n$ of E, then the relation $\boldsymbol{e}_j \wedge x = 0$ tells us that in the expansion $(*)$ there are no terms not containing $\boldsymbol{e}_j$. Therefore, we only have terms containing $\boldsymbol{e}_j$. Since j is arbitrary, the only possible non-zero term is the pseudoscalar $\lambda_N \boldsymbol{e}_N$, where $N = \{1, \ldots, n\}$.

E.2.7 Show that the expression (2.5) is alternating.

Hint We have seen that the cases $k = 2$ and $k = 3$ are obvious. Let us look again at the case $k = 3$,

$$i_{\boldsymbol{e}}(\boldsymbol{y}_1 \wedge \boldsymbol{y}_2 \wedge \boldsymbol{y}_3) = (\boldsymbol{e} \cdot \boldsymbol{y}_1)\boldsymbol{y}_2 \wedge \boldsymbol{y}_3 - (\boldsymbol{e} \cdot \boldsymbol{y}_2)\boldsymbol{y}_1 \wedge \boldsymbol{y}_3 + (\boldsymbol{e} \cdot \boldsymbol{y}_3)\boldsymbol{y}_1 \wedge \boldsymbol{y}_2.$$

If we swap $\boldsymbol{y}_1$ and $\boldsymbol{y}_2$, the second summand becomes $-(\boldsymbol{e} \cdot \boldsymbol{y}_1)\boldsymbol{y}_2 \wedge \boldsymbol{y}_3$; the first becomes $(\boldsymbol{e} \cdot \boldsymbol{y}_2)\boldsymbol{y}_1 \wedge \boldsymbol{y}_3 = -(-(\boldsymbol{e} \cdot \boldsymbol{y}_2)\boldsymbol{y}_1 \wedge \boldsymbol{y}_3)$; the third, $(\boldsymbol{e} \cdot \boldsymbol{y}_3)\boldsymbol{y}_2 \wedge \boldsymbol{y}_1 = -(\boldsymbol{e} \cdot \boldsymbol{y}_3)\boldsymbol{y}_1 \wedge \boldsymbol{y}_2$. A similar reasoning works when we swap $\boldsymbol{y}_2$ and $\boldsymbol{y}_3$, or, more generally, when we swap $\boldsymbol{y}_j$ and $\boldsymbol{y}_{j+1}$ $(1 \leqslant j < k)$.

E.2.8 If $d : \wedge E \to \wedge E$ is a graded skew-derivation of grade -1, then $d^2 = 0$.

Hint Check that $\widehat{dx} = -d\hat{x}$ and use it to get that $d^2(x \wedge x') = (d^2 x) \wedge x' + x \wedge (d^2 x')$. Finally, deduce that $d^2 = 0$ using this relation and the fact that d^2 vanishes on vectors.

E.2.9 (Degenerate Volumes Vanish) Show that a Gram determinant $G(\boldsymbol{x}_1, \ldots, \boldsymbol{x}_k)$ vanishes if $\boldsymbol{x}_1, \ldots, \boldsymbol{x}_k$ are linearly dependent.

Hint Argue that it suffices to establish the claim when $\boldsymbol{x}_k \in \langle \boldsymbol{x}_1, \ldots, \boldsymbol{x}_{k-1} \rangle$, and that in this case the last column of G is a linear combination of its first $k-1$ columns.

E.2.10 (Signature of $\wedge E$) Show that if $s > 0$, then the signature of $\wedge E_{r,s}$ is $(2^{n-1}, 2^{n-1})$.

Hint Use an orthogonal basis $\boldsymbol{e}_1, \ldots, \boldsymbol{e}_n$ of $E_{r,s}$ with $q(\boldsymbol{e}_j) > 0$ for $j = 1, \ldots, r$ and $q(\boldsymbol{e}_j) < 0$ for $j = r+1, \ldots, r+s = n$. With this, we have that $q(\boldsymbol{e}_{\hat{J}}) = q(\boldsymbol{e}_{j_1}) \cdots q(\boldsymbol{e}_{j_k})$, which is positive if and only if the cardinal of

$$J \cap \{r+1, \ldots, r+s = n\}$$

is even. This means that we can write the elements of the basis of $\wedge E$ in the form $\boldsymbol{e}_{\hat{J}} \wedge \boldsymbol{e}_{\hat{K}}$, with J a multiindex in $\{1, \ldots, r\}$ and K a multiindex in $\{r+1, \ldots, r+s\}$. Now, the claim follows because there are as many K with even cardinal as with odd cardinal, namely 2^{s-1} in each case (this is false precisely when $s = 0$), which implies that the signature of $\wedge E_{r,s}$ is $(2^r 2^{s-1}, 2^r 2^{s-1}) = (2^{n-1}, 2^{n-1})$.

E.2.11 Prove the congruences needed to complete the proof of 2.4.5.

Hint What has to be shown is that for positive integers k, l the integers $(m'-m)/\!\!/2$ and $l/\!\!/2 + k/\!\!/2 + kl + m$ have the same parity, where $m = \min(k,l)$ and $m' = \max(k,l)$. Since this relation does not change if we swap k and l, and it is true for $k = l$, we can assume that $k < l$. Now, consider a couple of cases that are sufficient to illustrate the reasoning. Suppose, for instance, that $k = 4k'+2$, $l = 4l'+3$, $k' \leqslant l'$. Then, $(m'-m)/\!\!/2 = (4(l'-k')+1)/\!\!/2 = 2(l'-k')$, which is even, and

$$l/\!\!/2 + k/\!\!/2 + kl + m = 2l' + 1 + 2k' + 1 + kl + k,$$

which is also even. For another instance, suppose that $k = 4k'+2$, $l = 4l'$, $k' < l'$. Then, $(m'-m)/\!\!/2 = (4(l'-k')-2)/\!\!/2 = 2(l'-k')-1$, which is odd, and

$$l/\!\!/2 + k/\!\!/2 + kl + m = 2l' + 1 + 2k' + kl + l,$$

which is also odd.

E.2.12 (Examples of the Laplace Rule) Let $\boldsymbol{x}_1, \boldsymbol{x}_2, \boldsymbol{y}_1, \boldsymbol{y}_2, \boldsymbol{y}_3 \in E$.

(1) Deduce the formula:

$$(\boldsymbol{x}_1 \wedge \boldsymbol{x}_2) \cdot (\boldsymbol{y}_1 \wedge \boldsymbol{y}_2) = (\boldsymbol{x}_1 \cdot \boldsymbol{y}_2)(\boldsymbol{x}_2 \cdot \boldsymbol{y}_1) - (\boldsymbol{x}_1 \cdot \boldsymbol{y}_1)(\boldsymbol{x}_2 \cdot \boldsymbol{y}_2)$$

in three ways: computing $q(\boldsymbol{x}_1 \wedge \boldsymbol{x}_2, \boldsymbol{y}_1 \wedge \boldsymbol{y}_2)$ (metric formula 2.4.7); using the first recursive formula (2.13); and using the second recursive formula (2.13).

(2) Compute the inner product $(\boldsymbol{x}_1 \wedge \boldsymbol{x}_2) \cdot (\boldsymbol{y}_1 \wedge \boldsymbol{y}_2 \wedge \boldsymbol{y}_3)$ in two ways: by using the Laplace rule and by developing it with the first recursive rule (2.13).

E.2.13 (About the Table of the Inner Product) Let us indicate how to get the second formula of Eq. (2.15) from the first and the commutation rule of the inner product. With the notations of that equation, we have $k \geqslant l$ and therefore

$$\boldsymbol{e}_{\hat{K}} \cdot \boldsymbol{e}_{\hat{L}} = (-1)^{kl+l} \boldsymbol{e}_{\hat{L}} \cdot \boldsymbol{e}_{\hat{K}}.$$

This vanishes unless $L \subseteq K$, and in this case it is equal to

$$(-1)^{kl+l}(-1)^{t(L,K)} q(\boldsymbol{e}_{\hat{L}}) \boldsymbol{e}_{(K-L)^{\hat{}}}.$$

But, $kl = t(K, L) + t(L, K) + l$, and so the sign is in fact $(-1)^{t(K,L)}$.
The same result can be obtained by using 2.4.5.

Chapter 3
Geometric Algebra

The aim of this chapter is to introduce the geometric algebra of a quadratic space (E, q) by following an axiomatic approach. The root idea is to explore how to *minimally* enrich the structure (E, q) so that vectors can be multiplied with the usual rules of an algebra, and that non-isotropic vectors can be inverted.

The main advantage of this approach is that it allows a balanced discussion of the similarities and differences of the various formalisms advocated by leading authors in the course of time. This is facilitated by the fact that the axiomatic method structures the flow of statements to be proved and provides strong hints about how to prove them. The approach also sheds light on the relative merits of different choices in the definition of primitive concepts and the corresponding notation systems. We will stick to those that seem most appropriate to us and comment alternatives with remarks along the way and, more systematically, in Sect. 6.3.

The axioms are stated in Sect. 3.1, and their significance is discussed. The notion of full geometric algebra appears in a natural way and it is proved that for each signature there exists a unique full geometric algebra up to isomorphism (Sect. 3.2). We also prove that non-full geometric algebras (which we call folded geometric algebras) can only occur for some special signatures, leaving the details of their construction to exercises. Section 3.3 is devoted to explain three main features of a full geometric algebra: its natural linear grading (here called Grassmann's grading), the outer (or exterior) product, and the inner (or interior) product. An important aspect in this context is the study of the most significant relations between these three concepts. The distinctive notion of pseudoscalars and the related (Hodge) duality are introduced in Sect. 3.4 and their fundamental properties established.

S. Xambó-Descamps, *Real Spinorial Groups*, SpringerBriefs in Mathematics,
https://doi.org/10.1007/978-3-030-00404-0_3

3.1 Axioms

Let us begin by an axiomatic presentation of the notion of *geometric algebra* which will allow us to find out its associated quadratic space, and in particular its signature, to settle existence and uniqueness results, and finally to lay open, for each signature (r, s), the rich structure of the unique full geometric algebra $\mathcal{G} = \mathcal{G}_{r,s}$ of that signature.
A *geometric algebra* (GA) is a structure featuring the ingredients described in **A0** and which satisfies the properties **A1–A4** that are stated below.

A0. *Structure*: An algebra $\mathcal{A}$ with a distinguished vector subspace $E \subseteq \mathcal{A}$ that does not contain the unit $1 = 1_{\mathcal{A}}$. The elements of $\mathbb{R} \subseteq \mathcal{A}$ (via the map $\lambda \mapsto \lambda 1_{\mathcal{A}}$) are called *scalars* and the elements of E, *vectors*. Note that $\mathbb{R} \cap E = \{0\}$. The product of $\mathcal{A}$ is called *geometric product* and is denoted by xy (juxtaposition of the factors x and y).

A1. *Minimality*: $\mathcal{A}$ is generated by E as an $\mathbb{R}$-algebra. In other words, $1_{\mathcal{A}}$ and the products of finite numbers of vectors span $\mathcal{A}$ as a vector space.

A2. *Contraction rule*: $\boldsymbol{x}^2 \in \mathbb{R}$ for any vector $\boldsymbol{x}$.
This axiom is essentially due to Clifford and has a fundamental character by comparison to the others, whose nature is basically technical.

The *magnitude* $|\boldsymbol{x}| \geqslant 0$ of $\boldsymbol{x} \in E$ is defined by the relation

$$|\boldsymbol{x}|^2 = \epsilon_{\boldsymbol{x}} \boldsymbol{x}^2, \tag{3.1}$$

where $\epsilon_{\boldsymbol{x}}$ denotes the sign of $\boldsymbol{x}^2$, which in this context is called the *signature* of $\boldsymbol{x}$. In particular we have that $|\boldsymbol{x}| = 0$ if and only if $\boldsymbol{x}^2 = 0$. Such vectors are said to be *null* or *isotropic*. The vectors of magnitude 1 are said to be *unit vectors*.

Notice that *if $\boldsymbol{x}$ is not null, then $\boldsymbol{x}$ is invertible* and

$$\boldsymbol{x}^{-1} = \boldsymbol{x}/\boldsymbol{x}^2 \in \langle \boldsymbol{x} \rangle \subseteq E. \tag{3.2}$$

If $\boldsymbol{x}, \boldsymbol{y} \in E$, we define $q(\boldsymbol{x}, \boldsymbol{y}) = \frac{1}{2}(\boldsymbol{x}\boldsymbol{y} + \boldsymbol{y}\boldsymbol{x})$.

3.1.1 *For all $\boldsymbol{x}, \boldsymbol{y} \in E$, $q(\boldsymbol{x}, \boldsymbol{y}) \in \mathbb{R}$.*

Proof The distributive property of the product allows us to write:

$$(\boldsymbol{x} + \boldsymbol{y})^2 = \boldsymbol{x}^2 + \boldsymbol{x}\boldsymbol{y} + \boldsymbol{y}\boldsymbol{x} + \boldsymbol{y}^2.$$

Since $\boldsymbol{x}^2, \boldsymbol{y}^2, (\boldsymbol{x} + \boldsymbol{y})^2 \in \mathbb{R}$, it follows that $\boldsymbol{x}\boldsymbol{y} + \boldsymbol{y}\boldsymbol{x} = 2q(\boldsymbol{x}, \boldsymbol{y}) \in \mathbb{R}$. □

Given that $q(\boldsymbol{x}, \boldsymbol{y}) \in \mathbb{R}$ is bilinear and symmetric as a function of $\boldsymbol{x}$ and $\boldsymbol{y}$, it is a metric (possibly degenerate) of E (*Clifford's metric*, or simply *metric* of $\mathcal{A}$). The expression

$$\boldsymbol{x}\boldsymbol{y} + \boldsymbol{y}\boldsymbol{x} = 2q(\boldsymbol{x}, \boldsymbol{y}) \tag{3.3}$$

will be called *Clifford's relation*. Setting $\boldsymbol{y} = \boldsymbol{x}$, we get $q(\boldsymbol{x}) = \boldsymbol{x}^2$, which means that the contraction rule is in fact equivalent to Clifford's relation.

Note that Clifford's relation implies that *two vectors $\boldsymbol{x}, \boldsymbol{y} \in E$ are q-orthogonal if and only if they anticommute, $\boldsymbol{x}\boldsymbol{y} = -\boldsymbol{y}\boldsymbol{x}$.*

A3. *Non-degeneracy*: We assume that the Clifford's metric q is non-degenerate. Its signature will be denoted (r, s), and we also say that it is the signature of $\mathcal{A}$.

For the formulation of **A4**, we have to make a detour for some preliminary considerations. They will be worth the effort, for they play an important role in the whole system.

Take a q-orthogonal basis $\mathbf{e}$ of E. If $n = \dim(E)$, let $N = \{1, \ldots, n\}$ (the set of *indices*; its subsets are called *multiindices*). If $K = k_1, \ldots, k_l \in N$ is a (finite) sequence of indices, we write $\boldsymbol{e}_K = \boldsymbol{e}_{k_1} \cdots \boldsymbol{e}_{k_l}$. Finally, let $B = \{\boldsymbol{e}_J\}$, $J \subseteq N$.

3.1.2 *The set B spans $\mathcal{A}$ as a vector space. In particular,* $\dim \mathcal{A} \leqslant 2^n$.

Proof The elements of the form $\boldsymbol{e}_K \in \mathcal{A}$, where K is any finite sequence of indices, span $\mathcal{A}$ as a vector space. This claim follows from **A1**, the bilinearity of the geometric product, and the convention that $\boldsymbol{e}_\emptyset = 1$.

A further reduction is that the $\boldsymbol{e}_K$ with $k_1 \leqslant \cdots \leqslant k_l$ also span $\mathcal{A}$ as a vector space. Indeed, since $\boldsymbol{e}_k \boldsymbol{e}_j = -\boldsymbol{e}_j \boldsymbol{e}_k$, the product $\boldsymbol{e}_K$ is equal to $(-1)^{t(K)} \boldsymbol{e}_{\tilde{K}}$, where $\tilde{K}$ is the result of reordering K in non-decreasing order and $t(K)$ is the number of order inversions in the sequence K.

After this reordering, the repeated factors in $\boldsymbol{e}_K$ appear grouped together and can be simplified using the contraction rule. The result is a scalar multiple of some $\boldsymbol{e}_J \in B$, and this completes the proof. □

3.1.3 (Artin's Formula [6]) *If I, J are multiindices, then*

$$\boldsymbol{e}_I \boldsymbol{e}_J = (-1)^{t(I,J)} q_{I \cap J}\, \boldsymbol{e}_{I \,\triangle\, J},$$

where $I \triangle J$ is the symmetric difference *of I and J, and*

$$q_K = q(\boldsymbol{e}_{k_1}) \cdots q(\boldsymbol{e}_{k_l})$$

for any index sequence $K = k_1, \ldots, k_l$. In particular,

$$\boldsymbol{e}_J^2 = (-1)^{|J| /\!/ 2} q_J.$$

Proof On reordering the product $\boldsymbol{e}_I \boldsymbol{e}_J$ in non-decreasing order of the indices, there occur $t(I, J)$ sign changes. Then the repeated factors have the form $e_k^2 = q(\boldsymbol{e}_k)$ for $k \in I \cap J$, and the remaining product is $\boldsymbol{e}_{I \,\triangle\, J}$.

For the second claim, it is enough to remember that

$$t(J, J) = (-1)^{\binom{|J|}{2}} = (-1)^{|J|//2}.$$

□

3.1.4 (Clifford Group of an Orthonormal Basis) If the basis **e** is orthonormal *(so $q_j = \pm 1$ for $j = 1, \ldots, n$, or, more generally, $q_J = \pm 1$* for any multiindex J*)*, then Artin's formula implies that $B^{\pm} = \{\pm \boldsymbol{e}_J \mid \boldsymbol{e}_J \in B\}$ is a multiplicative group. Note that $|B^{\pm}| \leqslant 2 \times 2^n = 2^{n+1}$, and that equality cannot be claimed because we cannot discard repetitions among the elements $\pm \boldsymbol{e}_J$ when J runs over all multiindices. □

3.1.5 (Commutation Formula) *If J and K are multiindices, then*

$$\boldsymbol{e}_K \boldsymbol{e}_J = (-1)^c (-1)^{|J| \times |K|} \boldsymbol{e}_J \boldsymbol{e}_K,$$

where $c = |I \cap J|$.

Proof Follows from a simple counting argument and Artin's formula. There are $|J| \times |K|$ pairs (j, k) $(j \in J, k \in K)$. The number of these pairs such that $j > k$ is $t(J, K)$; similarly, the number of pairs such that $k > j$ is $t(K, J)$; and there are c pairs such that $j = k$ (coincidences). Thus $|J| \times |K| = t(J, K) + t(K, J) + c$, which implies that $t(K, J) \equiv |J| \times |K| + c + t(J, K) \bmod 2$. The assertion is now an immediate consequence of Artin's formula, for $J \cap K = K \cap J$ and $J \,\triangle\, K = K \,\triangle\, J$.

□

3.1.6 (Effective Computations) The formulas 3.1.3 and 3.1.5 can be used to determine the table of geometric products of the generators B, which in turn can be used to calculate geometric products. As an illustration, see Exercises E.3.1, p. 56, and E.3.2, p. 57. Note that if the basis is orthonormal, then $\boldsymbol{e}_I \boldsymbol{e}_J = \pm \boldsymbol{e}_{I \triangle J}$, $\boldsymbol{e}_J \boldsymbol{e}_I = \pm \boldsymbol{e}_I \boldsymbol{e}_J$, and $\boldsymbol{e}_J^2 = \pm 1$, and that the signs can be readily determined with Artin's formula or the commutation formula. □

The analysis of the linear (in)dependence of B is usually formulated along the guidelines outlined by Riesz in [122], with amendments due to several authors (see, for example, the discussion in [128, pp. 43–44]). But unfortunately none of the amendments we know points to a residual implicit assumption in the basic argument. We will point out to this assumption in the presentation below, based on formulating axiom **A4** and deducing from it that the set B is linearly independent and hence that $\dim \mathcal{A} = 2^n$. That the axiom **A4** is indeed needed will be analyzed in 3.2.10.

A4. There exists an orthogonal basis **e** such that $\boldsymbol{e}_N \notin \mathbb{R}$.

3.1.7 (Remarks) If $\mathbf{e}'$ is another orthogonal basis, then $\boldsymbol{e}'_N \sim \boldsymbol{e}_N$ and hence **A4** is independent of which orthogonal basis we use. If in addition **e** is orthonormal, then the condition $\boldsymbol{e}_N \notin \mathbb{R}$ is equivalent to say that $\boldsymbol{e}_N \neq \pm 1$. □

In what follows, and until further notice, we assume **A4** and that $\mathbf{e}$ is any orthogonal basis of E. The implicit assumption to which we alluded above coincides with part (2) of the statement below.

3.1.8

(1) *For any non-empty* $L \subseteq N$, $\boldsymbol{e}_L \notin \mathbb{R}$.
(2) *If* $J, K \subseteq N$ *and* $J \neq K$, *then we have* $\boldsymbol{e}_J \boldsymbol{e}_K^{-1} \notin \mathbb{R}$ *or, equivalently,* $\boldsymbol{e}_J \not\sim \boldsymbol{e}_K$.

Proof

(1) We will argue that there does not exist a non-empty L such that $\boldsymbol{e}_L \in \mathbb{R}$. Indeed, this condition implies that $\boldsymbol{e}_L$ commutes with $\boldsymbol{e}_J$ for any J. By the commutation formula, this is equivalent to say that $|L \cap J| + |L||J|$ is even for any J. Taking $J = \{l\}$ for some $l \in L$ (which exists because L is non-empty), we see that $1 + |L|$ must be even and so $|L|$ must be odd. On the other hand, $L \neq N$, because $\boldsymbol{e}_N \notin \mathbb{R}$ by the axiom **A4**, so there is an index $j \in N - L$. Taking $J = \{j\}$ we get that $|L|$ must be even, which is a contradiction.
(2) We have $\boldsymbol{e}_K^{-1} \sim \boldsymbol{e}_K$ and $\boldsymbol{e}_J \boldsymbol{e}_K^{-1} \sim \boldsymbol{e}_J \boldsymbol{e}_K \sim \boldsymbol{e}_{J \,\triangle\, K} \notin \mathbb{R}$ (because $J \triangle K \neq \emptyset$).

□

3.1.9 (Riesz [122]) *The set B is linearly independent. Hence* $\dim \mathcal{A} = 2^n$.

Proof Assume that we have a linear relation $\sum_J \lambda_J \boldsymbol{e}_J = 0$. We want to show that then $\lambda_K = 0$ for any K. To that end, it suffices to see that $\lambda_\emptyset = 0$, for 3.1.8 guarantees that if we multiply the initial relation by $\boldsymbol{e}_K^{-1}$ then we obtain a similar relation in which the coefficient of $1 = \boldsymbol{e}_\emptyset$ is λ_K.

For each index k, the initial relation implies that $\sum_J \lambda_J \boldsymbol{e}_k \boldsymbol{e}_J \boldsymbol{e}_k^{-1} = 0$. Given that $\boldsymbol{e}_k$ commutes or anticommutes with $\boldsymbol{e}_J$, it is immediate to infer the relation $\sum_J \lambda_J \boldsymbol{e}_J = 0$ in which the sum is extended to all J such that $\boldsymbol{e}_J$ commutes with all $\boldsymbol{e}_k$. In this sum, the J such that $|J|$ is even and positive do not appear, for in this case $\boldsymbol{e}_J$ anticommutes with any of its factors. Nor can appear in it any J such that $|J|$ is odd and $J \neq N$, for in this case $\boldsymbol{e}_k$ anticommutes with $\boldsymbol{e}_J$ for any $k \in N - J$. We are therefore left with either the relation $\lambda_\emptyset = 0$ when n is even (in this case $J \neq N$ is automatic if $|J|$ is odd), or possibly the relation $\lambda_\emptyset + \lambda_N \boldsymbol{e}_N = 0$ if n is odd. But in this relation we must have $\lambda_N = 0$, for otherwise we would have $\boldsymbol{e}_N \in \mathbb{R}$. □

3.1.10 (Remark on A4) It is clear that if $\dim(\mathcal{A}) = 2^n$, then B is linearly independent for any orthogonal basis and hence **A4** is satisfied. This means that instead of **A4** we could have opted to assume directly that $\dim(\mathcal{A}) = 2^n$ (this is expressed by saying that the geometric algebra is *full*). This would have avoided 3.1.9, but it would leave us in the dark about the existence and nature of algebras satisfying all axioms except **A4**. In this brief these algebras will be said to be *folded* and we will describe them in detail in 3.2.10. In particular we will see that the most interesting low dimensional geometric algebras satisfy **A4** and therefore that they are automatically full. Nevertheless, folded algebras are not the oddity that they might seem, as they will show up in Sect. 6.3 to help in the classification of geometric algebras. We will also see that a folded geometric algebra on E_n is a full geometric algebra on a vector space E_{n-1}. □

3.1.11 (Functoriality and Uniqueness) *Let* $(\mathcal{A}, E)$ *and* $(\mathcal{A}', E')$ *be geometric algebras with the same signature, and assume that* $\mathcal{A}$ *is full. Let* $f : E \to E'$ *be an isometry. Then there exists a unique algebra homomorphism* $f^\sharp : \mathcal{A} \to \mathcal{A}'$ *whose restriction to E is f, i.e.,* $f^\sharp(\boldsymbol{x}) = f(\boldsymbol{x})$ *for all* $\boldsymbol{x} \in E$*. If* $\mathcal{A}'$ *is also full, then the homomorphism* $f^\sharp$ *is in fact an isomorphism.*

Proof Pick an orthonormal basis $\mathbf{e}$ of E and define $\mathbf{e}' = \boldsymbol{e}'_1, \ldots, \boldsymbol{e}'_n$, where $\boldsymbol{e}'_k = f(\boldsymbol{e}_k)$. Since f is an isometry, $\mathbf{e}'$ is an orthonormal basis of E'. Therefore, $B = \{\boldsymbol{e}_J\}$ is a basis of $\mathcal{A}$ and $B' = \{\boldsymbol{e}'_J\}$ is a linearly generating system for $\mathcal{A}'$. If $f^\sharp$ exists, we must have $f^\sharp(\boldsymbol{e}_J) = \boldsymbol{e}'_J$, and this shows that $f^\sharp$ is uniquely determined as a linear map, and that it is a linear isomorphism whenever $\mathcal{A}'$ is full.

To prove that $f^\sharp$ is a homomorphism of algebras, it is enough to see that $f(\boldsymbol{e}_J\boldsymbol{e}_K) = \boldsymbol{e}'_J\boldsymbol{e}'_K$ for all multiindices J and K. But this is an immediate consequence of Artin's formula, for if we set $L = J \triangle K$, then $\boldsymbol{e}_J\boldsymbol{e}_K = \epsilon \boldsymbol{e}_L$ and $\boldsymbol{e}'_J\boldsymbol{e}'_K = \epsilon \boldsymbol{e}'_L$, with the same sign ϵ. □

3.2 Existence of Full Geometric Algebras

In the context and notations of **A0** and **A2**, by the universal property of the tensor algebra there exists a unique algebra homomorphism $f : TE \to \mathcal{A}$ extending the inclusion $E \hookrightarrow \mathcal{A}$, so that

$$f(\boldsymbol{x}_1 \otimes \cdots \otimes \boldsymbol{x}_k) = \boldsymbol{x}_1 \cdots \boldsymbol{x}_k \text{ for all } \boldsymbol{x}_1, \ldots, \boldsymbol{x}_k \in E.$$

In particular we see that $f(\boldsymbol{x}\otimes\boldsymbol{x}) = \boldsymbol{x}^2 = q(\boldsymbol{x})$, and hence $\boldsymbol{x}\otimes\boldsymbol{x}-q(\boldsymbol{x}) \in \ker(f)$ for all $\boldsymbol{x} \in E$. It follows, using the universal property of quotients (see 1.4.6), that there exists a unique algebra homomorphism $\bar{f} : TE/I_qE \to \mathcal{A}$ such that $\bar{f}(\pi(\boldsymbol{x})) = \boldsymbol{x}$, where $\pi : TE \to TE/I_qE$ is the quotient of TE by the ideal I_qE generated by the elements of the form

$$\boldsymbol{x} \otimes \boldsymbol{x} - q(\boldsymbol{x}), \quad \boldsymbol{x} \in E. \tag{3.4}$$

This argument shows that the algebra $C_qE = TE/I_qE$, which is called the *Clifford algebra* of (E, q), is a candidate to be the full geometric algebra. And indeed it is, as we will argue in detail below. For example, the form of the generators (3.4) of I_qE is what will allow us to see that C_qE satisfies Clifford's contraction rule.

3.2.1 (Remark) Since the construction of the Clifford algebra C_qE and the properties 3.2.2–3.2.7 below are valid for a general metric q, presently we will not assume that q is regular until 3.2.8. Actually, an important special case is that the Clifford algebra of $q \equiv 0$ is the exterior algebra $\wedge E$ (see 2.1.8).

The Clifford algebra $C_q E$ is universal in the following sense (see E.3.3, p. 57): *for any algebra* $\mathcal{A}$ *and any linear map* $f : E \to \mathcal{A}$ *such that* $f(\boldsymbol{x})^2 = q(\boldsymbol{x})$ *for all* $\boldsymbol{x} \in E$, *there is a unique algebra homomorphism* $\bar{f} : C_q E \to \mathcal{A}$ *such that* $\bar{f}(\pi(\boldsymbol{x})) = f(\boldsymbol{x})$ *for any* $\boldsymbol{x} \in E$, *where* $\pi : TE \to C_q E$ *denotes the quotient map.*

3.2.2 (Involutions) The parity and reverse involutions of TE, $x \mapsto \hat{x}$ and $x \mapsto \tilde{x}$, leave invariant the ideal $I_q E$, and therefore they induce involutions of $C_q E$ that we will denote with the same symbols. If we write $\bar{x}$ to denote the image of $x \in TE$ in $C_q E$, the parity involution is the unique automorphism of $C_q E$ such that $\hat{\bar{\boldsymbol{x}}} = -\bar{\boldsymbol{x}}$ for all $\boldsymbol{x} \in E$, and the reverse involution is the unique antiautomorphism of $C_q E$ such that

$$(\bar{\boldsymbol{x}}_1 \cdots \bar{\boldsymbol{x}}_k)^\sim = \bar{\boldsymbol{x}}_k \cdots \bar{\boldsymbol{x}}_1 \tag{3.5}$$

for all $k \geqslant 1$ and any $\boldsymbol{x}_1, \ldots, \boldsymbol{x}_k \in E$.

3.2.3 (Contractions of $C_q E$) Given a linear form $\boldsymbol{\xi} \in E^*$, the contraction skew-derivation $i_\xi : TE \to TE$ (see 1.4.17) leaves invariant the ideal $I_q E$, because for any $\boldsymbol{x} \in E$ we have

$$i_\xi(\boldsymbol{x} \otimes \boldsymbol{x} - q(\boldsymbol{x})) = \xi(\boldsymbol{x})\boldsymbol{x} - \boldsymbol{x}\xi(\boldsymbol{x}) - i_\xi(q(\boldsymbol{x})) = 0.$$

It follows that i_ξ induces a linear endomorphism of $C_q E$ (still denoted i_ξ) defined by the formula

$$i_\xi(\bar{x}) = \overline{i_\xi x}.$$

This endomorphism is a skew-derivation of $C_q E$ in the sense that for all $x, y \in C_q E$ we have $i_\xi(xy) = (i_\xi x)y + \hat{x}(i_\xi y)$. As in the case of the tensor and exterior algebras, the skew-derivation i_x ($\boldsymbol{x} \in E$) is defined as i_{q_x}.

The next four statements establish, in an elementary way, basic properties of the Clifford algebra (in essence, they amount to checking that it satisfies the conditions **A0**). At times they are referred to concepts and constructions considerably more involved, and at others they are simply dispatched as obvious.

Given $\boldsymbol{x}, \boldsymbol{y} \in E$, we set $\tau(\boldsymbol{x}, \boldsymbol{y}) = \boldsymbol{x} \otimes \boldsymbol{y} + \boldsymbol{y} \otimes \boldsymbol{x} - 2q(\boldsymbol{x}, \boldsymbol{y}) \in TE$.

3.2.4 *If* $\boldsymbol{e}_1, \ldots, \boldsymbol{e}_n$ *is a basis of* E, $I_q E$ *is generated by* $\tau(\boldsymbol{e}_j, \boldsymbol{e}_k)$, $j \leqslant k$.

Proof The relation

$$\begin{aligned}(\boldsymbol{x} + \boldsymbol{y}) \otimes (\boldsymbol{x} + \boldsymbol{y}) - q(\boldsymbol{x} + \boldsymbol{y}) \\ = \boldsymbol{x} \otimes \boldsymbol{x} - q(\boldsymbol{x}) + \boldsymbol{y} \otimes \boldsymbol{y} - q(\boldsymbol{y}) + \boldsymbol{x} \otimes \boldsymbol{y} + \boldsymbol{y} \otimes \boldsymbol{x} - 2q(\boldsymbol{x}, \boldsymbol{y})\end{aligned}$$

shows that $\tau(\boldsymbol{x}, \boldsymbol{y}) \in I_q E$. Since $\tau(\boldsymbol{x}, \boldsymbol{x}) = 2(\boldsymbol{x} \otimes \boldsymbol{x} - q(\boldsymbol{x}))$, we see that the expressions $\tau(\boldsymbol{x}, \boldsymbol{y})$, for $\boldsymbol{x}, \boldsymbol{y} \in E$, generate $I_q E$. Now the claim follows from the bilinearity and symmetry of the function τ. □

3.2.5 $\bar{1} \neq \bar{0}$.

Proof If $\bar{1} = \bar{0}$, we would have $1 \in I_q E$. By 3.2.4, there exist $a_{jk}, b_{jk} \in TE$ such that $1 = \sum_{j \leqslant k} a_{jk} \tau(\boldsymbol{e}_j, \boldsymbol{e}_k) b_{jk}$. Since in this relation all components of positive grade vanish, we may assume that a_{jk} and b_{jk} are scalars. Now the grade 2 part of the relation gives us that $\sum_{j \leqslant k} a_{jk} b_{jk} (\boldsymbol{e}_j \otimes \boldsymbol{e}_k + \boldsymbol{e}_k \otimes \boldsymbol{e}_j) = 0$, and hence that $a_{jk} b_{jk} = 0$ for all $j \leqslant k$, because the elements $\boldsymbol{e}_j \otimes \boldsymbol{e}_k + \boldsymbol{e}_k \otimes \boldsymbol{e}_j$ $(1 \leqslant j \leqslant k \leqslant n)$ are linearly independent. But this leads, when we consider the grade 0 part, to $1 = -2 \sum_{j \leqslant k} a_{jk} b_{jk} q(\boldsymbol{e}_j, \boldsymbol{e}_k) = 0$, which is a contradiction. □

Thus we see that the map $\mathbb{R} \to C_q E$, $\lambda \mapsto \lambda \bar{1}$, is injective, and this allows us to identify $\mathbb{R}$ to the subfield $\mathbb{R}\bar{1}$ of $C_q E$.

3.2.6 *Let $\bar{E}$ be the image of E in $C_q E$. Then $\bar{1} \notin \bar{E}$.*

Proof If $\bar{1} = \bar{\boldsymbol{x}}$ for some $\boldsymbol{x} \in E$, the parity involution gives $\bar{1} = -\bar{\boldsymbol{x}}$ and from this we get the contradiction $\bar{1} = \bar{0}$. □

3.2.7 *The surjection $E \to \bar{E}$ is an isomorphism.*

Proof Let $\boldsymbol{x} \in E$ be such that $\bar{\boldsymbol{x}} = 0$. Then we have, for any linear form ξ,

$$\bar{0} = i_\xi(\bar{\boldsymbol{x}}) = \overline{i_\xi \boldsymbol{x}} = \overline{\xi(\boldsymbol{x})}$$

so $\xi(\boldsymbol{x}) = 0$ (by 3.2.5). Since ξ is arbitrary, we conclude that $\boldsymbol{x} = 0$. □

In what follows we will identify E with its image $\bar{E}$ in $C_q E$.

3.2.8 *If q is regular, $(C_q E, E)$ is a full geometric algebra with metric q.*

Proof The conditions **A0** have already been established (see 3.2.5–3.2.7).

The contraction rule **A2**, $\boldsymbol{x}^2 = q(\boldsymbol{x})$, follows from $\boldsymbol{x} \otimes \boldsymbol{x} - q(\boldsymbol{x}) \in I_q E$. This also shows that q is the metric of $C_q E$, so that **A3** is also satisfied.

Given that TE is generated by E as an $\mathbb{R}$-algebra, its quotient $C_q E$ has the same property, and so axiom **A1** also holds.

To end the proof we have to show that $C_q E$ is full, i.e., the $\boldsymbol{e}_N \notin \mathbb{R}$. Indeed, if $\boldsymbol{e}_N \in \mathbb{R}$, then n would be odd, because $\boldsymbol{e}_1$ (say) must commute with $\boldsymbol{e}_N$, and n would be even, because $\boldsymbol{e}_N$ is fixed by the parity involution. □

3.2.9 (On the Periodicity Theorem) The isomorphism class of the full geometric algebra $\mathcal{G}_{r,s}$ of signature (r, s) only depends on $n = r + s$ and $\nu = s - r \bmod 8$ and can be obtained explicitly. While the details about this order 8 *periodicity* are postponed until 6.3.3, the reader is encouraged to build in Exercises E.3.4–E.3.8 a few utilities that will be handy for that discussion.

3.2.10 (Folded Geometric Algebras) Let us use the notations introduced to formulate and discuss the axiom **A4** (p. 44).

Table 3.1 Special signatures for $n < 10$

n	3	5	7	9
(r, s)	(2,1), (0,3)	(5,0), (3,2), (1,4)	(6,1), (4,3), (2,5), (0,7)	(9,0), (7,2), (5,4), (3,6), (1,8)

If $(\mathcal{A}, E)$ is a folded geometric algebra, and $\mathbf{e}$ is an orthonormal basis of E, then we have $\boldsymbol{e}_N = \pm 1$. Now this condition implies that $\boldsymbol{e}_N$ commutes with vectors, and this can only happen if n is odd, say $n = 2m + 1$. On the other hand,

$$1 = \boldsymbol{e}_N^2 = (-1)^{n/\!/2+s} = (-1)^{m+s},$$

and so $m + s$ must be even. Hence s must be even (odd) if m is even (odd), which means that $n = 1 \bmod 4$ ($n = 3 \bmod 4$). Of such signatures we will say that they are *special* (see Table 3.1 for the special signatures up to $n = 9$).

For a non-special signature, there is a unique geometric algebra, which is full. Note that this happens for the Euclidean spaces whose dimension n is not congruent to 1 mod 4 (in particular for $n = 2, 3, 4$); for the *Lorentz signatures* $(n - 1, 1)$ for n not congruent to 3 mod 4 (in particular for $n = 4, 5, 6$, which includes the signature $(4, 1)$ of the five-dimensional conformal space of the Euclidean 3-space); and the *Minkowski signatures* $(1, n - 1)$ for n not congruent to 1 mod 4 (in particular for $n = 3, 4, 6, 7$, which includes the signature $(1, 3)$ of the four-dimensional ordinary Minkowski space).

For a special signature, besides the full geometric algebra there is a folded geometric algebra, unique up to isomorphism. It has dimension 2^{n-1} and can be constructed as the quotient $C_q E/(1 - \boldsymbol{e}_N)$, where $(1 - \boldsymbol{e}_N)$ is the ideal of $C_q E$ generated by $1 - \boldsymbol{e}_N$. For the details we refer to E.3.9, p. 61. □

3.3 The Grassmann's Grading and the Outer and Inner Products

For each signature (r, s) there exists a unique full geometric algebra, up to isomorphism. In fact, if $(\mathcal{A}, E)$ and $(\mathcal{A}', E')$ are full geometric algebras with the same signature, and $f : E \to E'$ is an isometry, then $f^\sharp : \mathcal{A} \to \mathcal{A}'$ is an algebra isomorphism (see 3.1.11). For the existence, see 3.2.8.

Thus we can speak about the (full) geometric algebra $\mathcal{G} = \mathcal{G}_{r,s}$ of signature (r, s). Recall that its product, the *geometric product*, is denoted by juxtaposition of its factors, $(x, y) \mapsto xy$, and that $B = \{\boldsymbol{e}_J\}$ denotes the linear basis of $\mathcal{G}$ associated with an orthogonal basis $\boldsymbol{e}_1, \ldots, \boldsymbol{e}_n$ of the vector space E of $\mathcal{G}$. We will also need the notation $B_k = \{\boldsymbol{e}_J : |J| = k\}$.

Consider the map $\mathsf{a} : E^k \to \mathcal{G}$ given by

$$\mathsf{a}(\boldsymbol{x}_1, \ldots, \boldsymbol{x}_k) = \tfrac{1}{k!}\textstyle\sum_J (-1)^{t(J)} \boldsymbol{x}_{j_1} \cdots \boldsymbol{x}_{j_k},$$

where the sum is extended to all permutations $J = j_1, \ldots, j_k$ of $\{1, \ldots, k\}$. This map is multilinear and skew-symmetric, and therefore *there exists a unique linear map*

$$\mathsf{g} : \wedge^k E \to \mathcal{G} \quad \text{such that} \quad \mathsf{g}(\boldsymbol{x}_1 \wedge \cdots \wedge \boldsymbol{x}_k) = \mathsf{a}(\boldsymbol{x}_1, \ldots, \boldsymbol{x}_k).$$

3.3.1 (The Subspace $\mathcal{G}^k$) *The map $\mathsf{g} : \wedge^k E \to \mathcal{G}$ is one-to-one and its image is the vector subspace $\mathcal{G}^k$ of $\mathcal{G}$ spanned by B_k. In particular, $\mathcal{G}^k$ is independent of the basis* **e** *used to define B and* $\dim \mathcal{G}^k = \binom{n}{k}$.

Proof If the vectors $\boldsymbol{x}_1, \ldots, \boldsymbol{x}_k$ are pair-wise orthogonal, then any two of them anticommute, which implies that $\mathsf{g}(\boldsymbol{x}_1 \wedge \cdots \wedge \boldsymbol{x}_k) = \boldsymbol{x}_1 \cdots \boldsymbol{x}_k$. In particular,

$$\mathsf{g}(\boldsymbol{e}_{\hat{J}}) = \boldsymbol{e}_J$$

for any multiindex J such that $|J| = k$. Since $\{\boldsymbol{e}_{\hat{J}} : |J| = k\}$ is a basis of $\wedge^k E$, the proof of all claims is immediate. □

Thus we have a *canonical linear grading*

$$\mathcal{G} = \mathcal{G}^0 \oplus \mathcal{G}^1 \oplus \cdots \oplus \mathcal{G}^n,$$

which we will call *Grassmann's grading*, and a canonical graded linear isomorphism $\mathsf{g} : \wedge E \to \mathcal{G}$. These claims just mean that any multivector x can be decomposed in a unique way as

$$x = x_0 + x_1 + \cdots + x_n, \quad x_k \in \mathcal{G}^k.$$

In fact, if $x = \sum_J \lambda_J \boldsymbol{e}_J$ is the expression of x as a linear combination of the basis B, then the grade k component of x is given by

$$x_k = \textstyle\sum_{|J|=k} \lambda_J \boldsymbol{e}_J.$$

Remark that many authors write $\langle x \rangle_k$ instead of x_k, and simply $\langle x \rangle$ instead $\langle x \rangle_0$.

Now the linear graded isomorphism $\mathsf{g} : \wedge E \to \mathcal{G}$ allows to graft onto $\mathcal{G}$ all notions and formalisms of $\wedge E$. Let us have a look at the more relevant ones.

3.3.2 (Outer Product) The outer or exterior product of $\wedge E$ produces an *outer* or *exterior product* in $\mathcal{G}$, $(x, y) \mapsto x \wedge y$. It is bilinear, unital, associative, graded, and skew-commutative. As established in the previous proof, the basic rules to manage this product are the following (with $\boldsymbol{x}_1, \ldots, \boldsymbol{x}_k, \boldsymbol{x}, \boldsymbol{y} \in E$):

1. $x_1 \wedge \cdots \wedge x_k = \mathsf{a}(x_1, \ldots, x_k)$. In particular, $x \wedge y = \mathsf{a}(x, y) = \frac{1}{2}(xy - yx)$.
2. If $x_1, \ldots, x_k$ are pair-wise orthogonal, $x_1 \wedge \cdots \wedge x_k = x_1 \cdots x_k$. In particular, $e_{\hat{J}} = e_J$ for any multiindex J whenever $\mathbf{e}$ is an orthogonal basis of E.
3. $e_J \wedge e_K = 0$ if $J \cap K \neq \emptyset$, and $e_J \wedge e_K = e_J e_K$ if $J \cap K = \emptyset$.

3.3.3 (Metric, Contraction Operators, and Inner Product) Next we can transfer the metric (still denoted q), the contraction operator, and the inner product. They are ruled by the laws explained in Sects. 2.2 and 2.4. In particular, the inner product is bilinear, but it is not associative and has no unit.

The following observation, which combines the geometric, inner and exterior products, can be considered as a birth certificate of geometric algebra. Its proof is obvious, because $x \cdot y = q(x, y) = \frac{1}{2}(xy + yx)$ and $x \wedge y = \frac{1}{2}(xy - yx)$.

$$xy = x \cdot y + x \wedge y. \tag{3.6}$$

We also can apply to $\mathcal{G}$ the terminology used for the exterior algebra. Thus the elements of $\mathcal{G}$ are called *multivectors*, and those of $\mathcal{G}^k$ are the *grade k* multivectors or *k-vectors*. Among the k-vectors there are the *k-blades*, which are non-zero k-vectors that are decomposable as a product $x_1 \wedge \cdots \wedge x_k$, $x_1, \ldots, x_k \in E$.

Other inheritances of $\mathcal{G}$ from $\wedge E$ are the parity and reverse involutions. They preserve the grading and will be denoted with the same symbols ($\hat{x}$ and $\tilde{x}$, respectively). For a k-vector, we have

$$\hat{x} = (-1)^k x \quad \text{and} \quad \tilde{x} = (-1)^{k/\!\!/2} x. \tag{3.7}$$

By 2.1.11 and 2.4.4, the parity involution is an automorphism of the exterior and inner products, while the reverse involution is, by 2.1.12 and 2.4.5, an antiautomorphism of both products.

Since the parity and reverse involutions of $\mathcal{G}$ coincide with the involutions of the same name of $C_q E$ introduced in 3.2.2, it turns out that the parity (reverse) involution is also an automorphism (antiautomorphism) of the geometric product:

$$\widehat{xy} = \hat{x}\hat{y}, \quad \widetilde{xy} = \tilde{y}\tilde{x}. \tag{3.8}$$

The contraction operators i_ξ (or i_x) of $\mathcal{G}$ introduced in 3.2.3 agree with those transferred from $\wedge E$ (see 2.2.1 for their definition and properties). In fact, i_ξ is a skew-derivation of $\mathcal{G}$ and this implies, using the definition of the outer product, that it satisfies Eq. (2.5), and the claim follows because both agree on vectors.

The multivectors that only contain even (odd) grade components are called *even* (*odd*) multivectors. The set of even (odd) multivectors is denoted by $\mathcal{G}^+$ ($\mathcal{G}^-$). Since $\mathcal{G}^+$ is closed by the geometric, exterior and inner products, it is called the *even geometric algebra*. The parity involution is the identity on $\mathcal{G}^+$, but the reversal involution induces an antiautomorphism of $\mathcal{G}^+$ which is the identity for grades of the form $4k'$ and minus the identity for grades of the form $4k' + 2$ (in particular for $k = 2$).

A Spread of Basic Formulas

To end the section, we collect a number of formulas that provide insights into the geometric algebra nature and facilitate theoretical and practical computations.

3.3.4 *For all* $\boldsymbol{e} \in E$ *and* $x \in \mathcal{G}$,

$$\boldsymbol{e}x = \boldsymbol{e} \cdot x + \boldsymbol{e} \wedge x = (i_{\boldsymbol{e}} + \mu_{\boldsymbol{e}})(x),$$

where $\mu_{\boldsymbol{e}} : \mathcal{G} \to \mathcal{G}$ *is the linear operator defined by the expression* $\mu_{\boldsymbol{e}}(x) = \boldsymbol{e} \wedge x$.

Proof Since both sides of the claimed equality are expressions that are bilinear in $\boldsymbol{e}$ and x, it is enough to check it for $\boldsymbol{e} = \boldsymbol{e}_k$ and $x = \boldsymbol{e}_J$, $k \in N$ and J an arbitrary multiindex. If $k \notin J$, then $\boldsymbol{e}_k \cdot \boldsymbol{e}_J = 0$ and $\boldsymbol{e}_k\boldsymbol{e}_J = \boldsymbol{e}_k \wedge \boldsymbol{e}_J$, and if $k \in J$, then $\boldsymbol{e}_k \wedge \boldsymbol{e}_J = 0$ and $\boldsymbol{e}_k\boldsymbol{e}_J = (-1)^{t(k,J)}q(\boldsymbol{e}_k)\boldsymbol{e}_{J-\{k\}} = \boldsymbol{e}_k \cdot \boldsymbol{e}_J$. Thus the relation is true in both cases. □

3.3.5 *For all* $\boldsymbol{e} \in E$ *and* $x \in \mathcal{G}$, $x\boldsymbol{e} = x \cdot \boldsymbol{e} + x \wedge \boldsymbol{e}$.

Proof $x\boldsymbol{e} = (\boldsymbol{e}\tilde{x})^{\sim} = (\boldsymbol{e} \cdot \tilde{x} + \boldsymbol{e} \wedge \tilde{x})^{\sim} = x \cdot \boldsymbol{e} + x \wedge \boldsymbol{e}$. □

3.3.6 (Riesz's Formulas) $2\boldsymbol{e} \wedge x = \boldsymbol{e}x + \hat{x}\,\boldsymbol{e}$ *and* $2\boldsymbol{e} \cdot x = \boldsymbol{e}x - \hat{x}\,\boldsymbol{e}$.

Proof The commutation rules of the inner and exterior product allow us to write $x \cdot \boldsymbol{e} = -\boldsymbol{e} \cdot \hat{x}$ and $x \wedge \boldsymbol{e} = \boldsymbol{e} \wedge \hat{x}$. Therefore we can write

$$x\,\boldsymbol{e} = -\boldsymbol{e} \cdot \hat{x} + \boldsymbol{e} \wedge \hat{x}.$$

Applied to $\hat{x}$, this identity yields

$$\hat{x}\,\boldsymbol{e} = -\boldsymbol{e} \cdot x + \boldsymbol{e} \wedge x.$$

Now this relation, together with $\boldsymbol{e}x = \boldsymbol{e} \cdot x + \boldsymbol{e} \wedge x$, easily yields the stated formulas. □

We will need the following corollary of Riesz's formulas:

3.3.7 *Let* x *be a multivector. If* $\boldsymbol{e}x = -x\boldsymbol{e}$ *for all* $\boldsymbol{e} \in E$, *then* x *is a pseudoscalar* (and n must be even).

Proof Let $x = x^+ + x^-$, where x^+ is even and x^- is odd. Then the condition says that $\boldsymbol{e}x^+ + \boldsymbol{e}x^- = -x^+\boldsymbol{e} - x^-\boldsymbol{e}$. In particular we must have, equating the parts of the same parity, the relations $\boldsymbol{e}x^+ = -x^+\boldsymbol{e}$ and $\boldsymbol{e}x^- = -x^-\boldsymbol{e}$. Now the latter is equivalent to $\boldsymbol{e}x^- - \widehat{x^-}\boldsymbol{e} = 0$, or $\boldsymbol{e} \cdot x^- = 0$ (by the second of Riesz's formulas). By 2.4.11, x^- must be a scalar, which is only possible, as scalars are even, if $x^- = 0$. So $x = x^+$ is even and we have $\boldsymbol{e}x + x\boldsymbol{e} = 0$. By the first of Riesz's formulas, we get that $\boldsymbol{e} \wedge x = 0$ for all vectors $\boldsymbol{e}$, and this implies that x is a pseudoscalar (see E.2.6, p. 38).

3.3.8 (Grades of a Geometric Product) *Let* $x \in \mathcal{G}^k$ *and* $y \in \mathcal{G}^l$*. If* $p \in \{0, 1, \ldots, n\}$ *and* $(xy)_p \neq 0$*, then* $p = |k - l| + 2i$ *with* $i \geqslant 0$ *and* $p \leqslant k + l$*. Moreover,* $(xy)_{k+l} = x \wedge y$ *and if* $k, l > 0$*,* $(xy)_{|k-l|} = x \cdot y$*.*

Proof Since $(xy)_p$ depends linearly on x, we may assume that x is a k-blade, say $x = X = \boldsymbol{x}_1 \wedge \cdots \wedge \boldsymbol{x}_k$, and that $\boldsymbol{x}_1, \ldots, \boldsymbol{x}_k$ is an orthogonal basis of $|X\rangle = \langle \boldsymbol{x}_1, \ldots, \boldsymbol{x}_k\rangle$. So $X = \boldsymbol{x}_1 \cdots \boldsymbol{x}_k$ and, using 3.3.4,

$$xy = (\mu_{\boldsymbol{x}_1} + i_{\boldsymbol{x}_1}) \cdots (\mu_{\boldsymbol{x}_k} + i_{\boldsymbol{x}_k})(y).$$

If in the development of the products on the right-hand side we choose i times the summand μ, and hence $k - i$ times the contraction operator, we get a multivector of grade $l + i - (k - i) = l - k + 2i$. The maximum grade that we can form in this way is $k + l$, obtained when $i = k$, and the corresponding term is $x \wedge y$.

If $0 < k \leqslant l$, the minimum grade in the development is $l - k$ (achieved with $i = 0$), and it is equal to

$$(i_{\boldsymbol{x}_1} \cdots i_{\boldsymbol{x}_k})(y) = i_x(y) = x \cdot y,$$

by Eqs. (2.8) and (2.12). And if $k \geqslant l > 0$, the minimum grade that appears in xy is the minimum grade that appears in $\widetilde{xy} = \tilde{y}\,\tilde{x}$, namely $k - l$, and the corresponding term is $((\tilde{y}\,\tilde{x})_{k-l})^{\sim} = (\tilde{y} \cdot \tilde{x})^{\sim} = x \cdot y$. □

3.3.9 (The Metric in Terms of the Geometric Product) *For all* $x, y \in \mathcal{G}$*, we have* $q(x, y) = (\tilde{x}y)_0 = (x\tilde{y})_0$*. In particular we have* $q(x) = (\tilde{x}x)_0 = (x\tilde{x})_0$ *for all* $x \in \mathcal{G}$*.*

Proof Since both expressions are bilinear, we can assume that x and y are homogeneous. Let k and l be their respective grades. Then 3.3.8 tells us that $(\tilde{x}y)_0 = (x\tilde{y})_0 = 0$ when $k \neq l$, which is the value of $q(x, y)$ in this case. So suppose that $k = l$. Then $(\tilde{x}y)_0 = \tilde{x} \cdot y$, again by 3.3.8, and $\tilde{x} \cdot y = q(x, y)$ by the metric formula 2.4.7. Finally, $q(x, y) = q(y, x) = \tilde{y} \cdot x = x \cdot \tilde{y} = (x\tilde{y})_0$. □

3.3.10 (Invertible Blades) *If* X *is a k-blade, then*

$$q(X) = \tilde{X}X = X\tilde{X} = (-1)^{k/\!\!/2}X^2.$$

In particular we see that X *is invertible if and only if* $X^2 \neq 0$*, or if and only if* $q(X) \neq 0$*, and if this is the case, then we have*

$$X^{-1} = X/X^2 = \tilde{X}/q(X).$$

Proof By the preceding result, it suffices to see that $\tilde{X}X$ is a scalar. For this we may assume that $X = \boldsymbol{x}_1 \cdots \boldsymbol{x}_k$, where $\boldsymbol{x}_1, \ldots, \boldsymbol{x}_k$ is an orthogonal basis of $|X\rangle$, and then $\tilde{X}X = \boldsymbol{x}_1^2 \cdots \boldsymbol{x}_k^2 \in \mathbb{R}$.

The last claimed expression is obvious because $\tilde{X} = (-1)^{k/\!\!/2}X$. □

3.3.11 (Remark) Together with 2.3.2, we see that the space $|X\rangle$ of a non-zero blade X is singular if and only if $X^2 = 0$.

Physically motivated authors often refer to the cases $X^2 > 0$, $X^2 < 0$, and $X^2 = 0$ as *time-like*, *space-like*, and *light-like*, respectively, thus extending the terminology used for the vectors of the Minkowski space $E_{1,3}$.

We could also label them, following an age-old terminology, as *elliptic*, *hyperbolic*, and *parabolic* (this option will be adopted for plane rotors in Sect. 5.2), or simply as *positive*, *negative*, and *null*, respectively.

3.4 Pseudoscalars

Let $\mathbf{e} = \boldsymbol{e}_1, \ldots, \boldsymbol{e}_n$ be an orthonormal basis of $E = E_{r,s}$ and define

$$I_{\mathbf{e}} = \boldsymbol{e}_1 \wedge \cdots \wedge \boldsymbol{e}_n \in \mathcal{G}^n.$$

We will say that $I_{\mathbf{e}}$ is the *pseudoscalar* associated with $\mathbf{e}$. Note that the metric formula gives us that

$$q(I_{\mathbf{e}}) = q(\boldsymbol{e}_1) \cdots q(\boldsymbol{e}_n) = (-1)^s.$$

If $\mathbf{e}' = \boldsymbol{e}'_1, \ldots, \boldsymbol{e}'_n$ is another orthonormal basis of E, then

$$I_{\mathbf{e}'} = \delta I_{\mathbf{e}},$$

where $\delta = \det_{\mathbf{e}}(\mathbf{e}')$ is the determinant of the matrix of the vectors $\mathbf{e}'$ with respect to the basis $\mathbf{e}$. Now the equalities

$$q(I_{\mathbf{e}}) = q(I_{\mathbf{e}'}) = q(\delta I_{\mathbf{e}}) = \delta^2 q(I_{\mathbf{e}})$$

allow us to conclude that $\delta = \pm 1$. This means that, up to sign, there is a unique pseudoscalar. The distinction of one of them amounts to choose an *orientation* for E.

3.4.1 (Properties of the Pseudoscalar) *Let $I \in \mathcal{G}^n$ be a pseudoscalar and $\mathcal{G}^\times$ the group of invertible multivectors with respect to the geometric product. Then*

(1) $I \in \mathcal{G}^\times$, $\quad I^{-1} = (-1)^s \tilde{I} = (-1)^s(-1)^{n//2} I$, $\quad I^2 = (-1)^{n//2}(-1)^s$.

(2) Hodge duality.

For any $x \in \mathcal{G}^k$, $Ix, xI \in \mathcal{G}^{n-k}$ and the maps $x \mapsto Ix$ and $x \mapsto xI$ are linear isomorphisms $\mathcal{G}^k \to \mathcal{G}^{n-k}$. The inverse maps are given by $x \mapsto I^{-1}x$ and $x \mapsto xI^{-1}$, respectively.

(3) *If n is odd, I commutes with all the elements of $\mathcal{G}$* (this is expressed by saying that I is a *central* element of $\mathcal{G}$) *. If n is even, I commutes (anticommutes) with even (odd) multivectors.*

(4) *If* $q(I) = 1$ $(q(I) = -1)$, *the Hodge duality maps are* isometries (antiisometries).

Proof

(1) Since $(-1)^s = q(I) = \tilde{I}\,I$, it is clear that $I \in \mathcal{G}^\times$ and that I^{-1} is given by the expression in the statement. Finally we have $I^2 = (-1)^{n/\!/2}\tilde{I}I = (-1)^{n/\!/2}(-1)^s$.
(2) Given that $I = \boldsymbol{e}_N$, Artin's formula 3.1.3 yields that $\boldsymbol{e}_J I, I\boldsymbol{e}_J \in \mathcal{G}^{n-k}$ for any multiindex J of grade k. All other claims are obvious.
(3) The commutation rule 3.1.5 gives

$$\boldsymbol{e}_j I = \boldsymbol{e}_j \boldsymbol{e}_N = (-1)^{n+1}\boldsymbol{e}_N\boldsymbol{e}_j = (-1)^{n+1}I\boldsymbol{e}_j,$$

and this shows that I commutes (anticommutes) with all vectors if n is odd (if n is even).
(4) If $x, y \in \mathcal{G}^k$, the alternative definition of the metric gives

$$q(xI, yI) = \left(xI\,\widetilde{yI}\,\right)_0 = (xI\,\tilde{I}\tilde{y})_0 = (xq(I)\tilde{y})_0 = q(I)q(x, y).$$

That $q(Ix, Iy) = q(I)q(x, y)$ is shown in a similar way:

$$\widetilde{Ix}\,Iy = \tilde{x}\tilde{I}I\tilde{y} = \tilde{x}q(I)y = q(I)\tilde{x}y.$$

□

3.4.2 (Example: Cross Product) In the case of (oriented) E_3, the Hodge duality

$$E_3 = \mathcal{G}^1 \to \mathcal{G}^2, \quad \boldsymbol{x} \mapsto \boldsymbol{x}^* = \boldsymbol{x}\boldsymbol{i} = \boldsymbol{i}\boldsymbol{x},$$

is an isometry, where in this case the pseudoscalar $\boldsymbol{e}_1\boldsymbol{e}_2\boldsymbol{e}_3$ ($\boldsymbol{e}_1, \boldsymbol{e}_2, \boldsymbol{e}_3$ is a positively oriented orthonormal basis) is denoted by $\boldsymbol{i}$. The inverse isometry is given by

$$b \mapsto b^* = -b\boldsymbol{i} = -\boldsymbol{i}b, \quad b \in \mathcal{G}^2.$$

Note that for any vector $\boldsymbol{x}$ we have

$$\boldsymbol{x} \wedge \boldsymbol{x}^* = \boldsymbol{x} \wedge (\boldsymbol{x}\boldsymbol{i}) = \boldsymbol{x} \wedge (\boldsymbol{x} \cdot \boldsymbol{i}) = \boldsymbol{x}^2\boldsymbol{i},$$

because $\boldsymbol{x} \wedge \boldsymbol{i} = 0$ and so $0 = \boldsymbol{x} \cdot (\boldsymbol{x} \wedge \boldsymbol{i}) = \boldsymbol{x}^2\boldsymbol{i} - \boldsymbol{x} \wedge (\boldsymbol{x} \cdot \boldsymbol{i})$.

Now the *cross product* $\boldsymbol{x} \times \boldsymbol{y}$ of two vectors $\boldsymbol{x}, \boldsymbol{y} \in E_3$ is defined as the Hodge dual of $\boldsymbol{x} \wedge \boldsymbol{y}$, namely

$$\boldsymbol{x} \times \boldsymbol{y} = (\boldsymbol{x} \wedge \boldsymbol{y})^* = -(\boldsymbol{x} \wedge \boldsymbol{y})\boldsymbol{i}.$$

Using notations from 2.3.3, the computation

$$|\boldsymbol{x} \times \boldsymbol{y}|^2 = (\boldsymbol{x} \times \boldsymbol{y})^2 = q(\boldsymbol{x} \times \boldsymbol{y}) = q(\boldsymbol{x} \wedge \boldsymbol{y}) = G(\boldsymbol{x}, \boldsymbol{y}) = A(\boldsymbol{x}, \boldsymbol{y})^2$$

shows that

$$|\boldsymbol{x} \wedge \boldsymbol{y}| = A(\boldsymbol{x}, \boldsymbol{y}),$$

where $A(\boldsymbol{x}, \boldsymbol{y})$ denotes the *area* (2-volume) of the parallelogram defined by $\boldsymbol{x}$ and $\boldsymbol{y}$.

Since $(\boldsymbol{x} \wedge \boldsymbol{y})\boldsymbol{i} = (\boldsymbol{x} \wedge \boldsymbol{y}) \cdot \boldsymbol{i}$, we immediately get that $\boldsymbol{x} \wedge \boldsymbol{y}$ is orthogonal to $\boldsymbol{x}$ and $\boldsymbol{y}$. For example,

$$\boldsymbol{x} \cdot (\boldsymbol{x} \times \boldsymbol{y}) = -\boldsymbol{x} \cdot ((\boldsymbol{x} \wedge \boldsymbol{y}) \cdot \boldsymbol{i}) = (\boldsymbol{x} \wedge \boldsymbol{x} \wedge \boldsymbol{y}) \cdot \boldsymbol{i} = 0.$$

In addition, the triple $\boldsymbol{x}, \boldsymbol{y}, \boldsymbol{x} \times \boldsymbol{y}$ is positively oriented, because

$$(\boldsymbol{x} \times \boldsymbol{y}) \wedge \boldsymbol{x} \wedge \boldsymbol{y} = (\boldsymbol{x} \times \boldsymbol{y}) \wedge (\boldsymbol{x} \times \boldsymbol{y})\boldsymbol{i} = (\boldsymbol{x} \times \boldsymbol{y})^2\boldsymbol{i} = A(\boldsymbol{x}, \boldsymbol{y})^2\boldsymbol{i}.$$

For practical computations the following instances may be helpful:

$$\boldsymbol{e}_1 \times \boldsymbol{e}_2 = \boldsymbol{e}_3, \quad \boldsymbol{e}_2 \times \boldsymbol{e}_3 = \boldsymbol{e}_1, \quad \boldsymbol{e}_3 \times \boldsymbol{e}_1 = \boldsymbol{e}_2$$

For example,

$$\boldsymbol{e}_1 \times \boldsymbol{e}_2 = -(\boldsymbol{e}_1 \wedge \boldsymbol{e}_2)\boldsymbol{i} = -(\boldsymbol{e}_1\boldsymbol{e}_2)\boldsymbol{e}_1\boldsymbol{e}_2\boldsymbol{e}_3 = \boldsymbol{e}_3.$$

3.4.3 (Remark) The equation $\boldsymbol{x} \wedge (\boldsymbol{x}\boldsymbol{i}) = \boldsymbol{x}^2\boldsymbol{i}$ used (and proved) in the example above is a special case of the following general formula. Let $\mathcal{G} = \mathcal{G}_{r,s}$ and I a pseudoscalar. If $x \in \mathcal{G}^k$, then $\tilde{x} \wedge (xI) = q(x)I$. This is clear if $x = \boldsymbol{e}_K$ ($\boldsymbol{e}_1, \ldots, \boldsymbol{e}_n$ an orthogonal basis), because $\tilde{\boldsymbol{e}}_K \wedge (\boldsymbol{e}_K I) = \tilde{\boldsymbol{e}}_K(\boldsymbol{e}_K I) = q(\boldsymbol{e}_K)I$. If $x = \sum_K \lambda_K \boldsymbol{e}_K$, then $xI = \sum_L \lambda_L(\boldsymbol{e}_L I)$ and $\tilde{x} \wedge (xI) = \sum_K \lambda_K^2 q(\boldsymbol{e}_K)I = q(x)I$. We have used that $\tilde{\boldsymbol{e}}_K \wedge (\boldsymbol{e}_L I) = 0$ for $L \neq K$, which is clear because in this case the blades $\tilde{\boldsymbol{e}}_K$ and $\boldsymbol{e}_L I$ share a factor $\boldsymbol{e}_i$ for any $i \in K - L$.

3.5 Exercises

E.3.1 Let $\mathcal{G}_2$ be the geometric algebra of the Euclidean plane E_2 and $\mathcal{G}_{\bar{2}}$ the geometric algebra of the anti-Euclidean plane $E_{\bar{2}}$ (its metric is $\bar{q} = -q$ if q is the metric of E_2).

Let $\boldsymbol{e}_1, \boldsymbol{e}_2$ be an *orthonormal* basis of E_2. The corresponding Clifford basis of $\mathcal{G}_2$ and of $\mathcal{G}_{\bar{2}}$ is the same, namely $1, \boldsymbol{e}_1, \boldsymbol{e}_2, \boldsymbol{e}_{12} = \boldsymbol{e}_1\boldsymbol{e}_2$, but the tables of the geometric product are quite different. Writing $\boldsymbol{i} = \boldsymbol{e}_1\boldsymbol{e}_2$ (the notation is suggested by the relation $\boldsymbol{i}^2 = -1$), we have:

$\mathcal{G}_2$	e_1	e_2	i
e_1	1	i	e_2
e_2	$-i$	1	$-e_1$
i	$-e_2$	e_1	-1

$\bar{\mathcal{G}}_2$	e_1	e_2	i
e_1	-1	i	$-e_2$
e_2	$-i$	-1	e_1
i	e_2	$-e_1$	-1

E.3.2 Let $\mathcal{G} = \mathcal{G}_3$ be the geometric algebra of the Euclidean space E_3. Let $\mathbf{e} = e_1, e_2, e_3$ be an *orthonormal basis* and set $i = e_{123} = e_1e_2e_3$ (the notation is again suggested by the relation $i^2 = -1$, which is immediate using Artin's formula). Check that $ie_1 = e_2e_3, ie_2 = e_3e_1, ie_3 = e_1e_2$ and hence that $1, e_1, e_2, e_3, ie_1, ie_2, ie_3, i$ is a basis of $\mathcal{G}$ (it agrees with the Clifford basis associated with $\mathbf{e}$ except for the order of the terms ie_k and the sign of ie_2). Check that the table of the geometric product using this basis is as follows:

$\mathcal{G}_3$	e_1	e_2	e_3	ie_1	ie_2	ie_3	i
e_1	1	ie_3	$-ie_2$	i	$-e_3$	e_2	ie_1
e_2	$-ie_3$	1	ie_1	e_3	i	$-e_1$	ie_2
e_3	ie_2	$-ie_1$	1	$-e_2$	e_1	i	ie_3
ie_1	i	$-e_3$	e_2	-1	$-ie_3$	ie_2	$-e_1$
ie_2	e_3	i	$-e_1$	ie_3	-1	$-ie_1$	$-e_2$
ie_3	$-e_2$	e_1	i	$-ie_2$	ie_1	-1	$-e_3$
i	ie_1	ie_2	ie_3	$-e_1$	$-e_2$	$-e_3$	-1

For example, $(ie_1)e_3 = (e_2e_3)e_3 = e_2e_3^2 = e_2$ and $e_3(ie_1) = e_3(e_2e_3) = -e_2e_3^2 = -e_2$. Note that the *complex scalars* $\mathbf{C} = \langle 1, i\rangle = \{\lambda + \mu i \mid \lambda, \mu \in \mathbb{R}\}$ form a field isomorphic to $\mathbb{C}$, $\boldsymbol{\xi} = \lambda + \mu i \mapsto \xi = \lambda + \mu i$, and that $\mathcal{G}_3$ is a $\mathbf{C}$-vector space with basis $1, e_1, e_2, e_3$, so that any element of $\mathcal{G}_3$ can be uniquely written in the form $\xi_0 + \xi_1e_1 + \xi_2e_2 + \xi_3e_3$, $\xi_0, \xi_1, \xi_2, \xi_3 \in \mathbf{C}$.

E.3.3 (Universal Property of the Clifford Algebra C_qE) Prove the universal property of C_qE as stated in Remark 3.2.1.

Hint By the universal property of the tensor algebra, there is a unique algebra homomorphism $\tilde{f} : TE \to \mathcal{A}$ such that $\tilde{f}(x) = f(x)$ for all $x \in E$, and this map satisfies $\tilde{f}(x \otimes x - q(x)) = 0$ for any $x \in E$. Now use the properties of the quotient $C_qE = TE/I_qE$ to construct $\bar{f}$.

The purpose of Exercises E.3.4–E.3.8 is to establish several isomorphisms. They include the determination of the isomorphism class of $\mathcal{G}_{r,s}$ for $1 \leqslant r + s \leqslant 3$ and they will be an effective tool for the determination of the isomorphism class of all geometric algebras in Sect. 6.3.

If $\mathcal{A}$ is any algebra and d a positive integer, we write $2\mathcal{A}$, or $\mathcal{A}+\mathcal{A}$, to denote the algebra of pairs $\{(a, a') \mid a, a' \in \mathcal{A}\}$ with the component-wise sum and the product $(a, a')(b, b') = (ab, a'b')$, and $\mathcal{A}(d)$ to denote the algebra of $d \times d$ matrices with entries in $\mathcal{A}$.

E.3.4 ($\mathcal{G}_1 \simeq \mathbb{R} \oplus \mathbb{R}$) Let $\boldsymbol{e} = \boldsymbol{e}_1$ be a unit vector of E_1. Show that the linear isomorphism $\mathcal{G}_1 \to 2\mathbb{R}$, $\alpha + \beta\boldsymbol{e} \mapsto (\alpha+\beta, \alpha-\beta)$ is an algebra isomorphism, and that the parity involution is given, in terms of $2\mathbb{R}$, by $(a,b)\hat{} = (b,a)$.

Hint Check that $u := \frac{1}{2}(1+\boldsymbol{e})$ and $u' := \frac{1}{2}(1-\boldsymbol{e})$ satisfy the relations

$$u + u' = 1, \quad u^2 = u, \quad u'^2 = u', \quad uu' = u'u = 0,$$

and that for all $\alpha + \beta\boldsymbol{e} \in \mathcal{G}^1$

$$(\alpha + \beta\boldsymbol{e}) = (\alpha+\beta\boldsymbol{e})u + (\alpha+\beta\boldsymbol{e})u' = (\alpha+\beta)u + (\alpha-\beta)u'.$$

E.3.5 ($\mathcal{G}_2 \simeq \mathbb{R}(2)$) Let $\boldsymbol{e}_1, \boldsymbol{e}_2$ be an orthonormal basis of E_2. Show that the linear isomorphism $\mathcal{G}_2 \to \mathbb{R}(2)$,

$$\alpha + \beta_1\boldsymbol{e}_1 + \beta_2\boldsymbol{e}_2 + \gamma\boldsymbol{e}_1\boldsymbol{e}_2 \mapsto \begin{pmatrix} \alpha+\beta_1 & \gamma+\beta_2 \\ -\gamma+\beta_2 & \alpha-\beta_1 \end{pmatrix}$$

is an isomorphism of algebras and that the parity and reverse involutions become, in terms of $\mathbb{R}(2)$, the cofactor and transpose operators:

$$\widehat{\begin{pmatrix} a & b \\ c & d \end{pmatrix}} = \begin{pmatrix} d & -c \\ -b & a \end{pmatrix}, \quad \widetilde{\begin{pmatrix} a & b \\ c & d \end{pmatrix}} = \begin{pmatrix} a & c \\ b & d \end{pmatrix}.$$

Note that the even subalgebra

$$\mathcal{G}_2^+ = \{\alpha + \gamma\boldsymbol{e}_1\boldsymbol{e}_2 \mid \alpha, \gamma \in \mathbb{R}\} \simeq \mathbb{C}$$

(as $(\boldsymbol{e}_1\boldsymbol{e}_2)^2 = -1$) is isomorphic to the subalgebra of real matrices $\begin{pmatrix} \alpha & \gamma \\ -\gamma & \alpha \end{pmatrix}$ and that the reversion involution for the subalgebra is morphed to the complex conjugation.

Hint Consider the linear map $m : E_2 \to \mathbb{R}(2)$, $\beta_1\boldsymbol{e}_1 + \beta_2\boldsymbol{e}_2 \mapsto \beta_1\mathfrak{e}_1 + \beta_2\mathfrak{e}_2$, where $\mathfrak{e}_1 = \begin{pmatrix} 1 & 0 \\ 0 & -1 \end{pmatrix}$ and $\mathfrak{e}_2 = \begin{pmatrix} 0 & 1 \\ 1 & 0 \end{pmatrix}$. Then we have $\mathfrak{e}_1^2 = \mathfrak{e}_2^2 = I_2$, $\mathfrak{e}_1\mathfrak{e}_2 = -\mathfrak{e}_2\mathfrak{e}_1$ and hence $m(\beta_1\boldsymbol{e}_1 + \beta_2\boldsymbol{e}_2)^2 = \beta_1^2 + \beta_2^2 = q(\beta_1\boldsymbol{e}_1 + \beta_2\boldsymbol{e}_2)$. By the universal property E.3.3, there is a unique algebra homomorphism $\mathcal{G}_2 \to \mathbb{R}(2)$ that extends m, and this homomorphism has the claimed form. For the second part, note that the parity involution changes the signs of β_1 and β_2, but not of α and γ, whereas the reverse involution changes the sign of γ, but not that of α, β_1, and β_2.

E.3.6 ($\mathcal{G}_{r+1,s+1} \simeq \mathcal{G}_{r,s}(2)$) Consider a decomposition

$$E_{r+1,s+1} = E_{r,s} \perp \langle \boldsymbol{e}, \bar{\boldsymbol{e}} \rangle,$$

where $\boldsymbol{e}^2 = -\bar{\boldsymbol{e}}^2 = 1$ and $\boldsymbol{e} \cdot \bar{\boldsymbol{e}} = 0$. Then any element of $\mathcal{G}_{r+1,s+1}$ can be written in a unique way in the form $a + b\boldsymbol{e} + \bar{b}\bar{\boldsymbol{e}} + c\boldsymbol{e}\bar{\boldsymbol{e}}$, $a, b, \bar{b}, c \in \mathcal{G}_{r,s}$. Show that the linear isomorphism $\mathcal{G}_{r+1,s+1} \to \mathcal{G}_{r,s}(2)$,

$$a + b\boldsymbol{e} + \bar{b}\bar{\boldsymbol{e}} + c\boldsymbol{e}\bar{\boldsymbol{e}} \mapsto \begin{pmatrix} a+b & c+\bar{b} \\ c-\bar{b} & a-b \end{pmatrix}$$

is an isomorphism of algebras. Show also that the parity involution becomes, in terms of matrices, the operator

$$\widehat{\begin{pmatrix} x & y \\ w & z \end{pmatrix}} = \begin{pmatrix} \hat{z} & \hat{w} \\ \hat{y} & \hat{x} \end{pmatrix}.$$

In particular we have $\mathcal{G}_{1,1} \simeq \mathbb{R}(2)$, by

$$\alpha + \beta\boldsymbol{e} + \bar{\beta}\bar{\boldsymbol{e}} + \gamma\boldsymbol{e}\bar{\boldsymbol{e}} \mapsto \begin{pmatrix} \alpha+\beta & \gamma+\bar{\beta} \\ \gamma-\bar{\beta} & \alpha-\beta \end{pmatrix}, \quad \alpha, \beta, \bar{\beta}, \gamma \in \mathbb{R}. \tag{3.9}$$

Hint Any element of $E_{r+1,s+1}$ can be written in a unique way in the form $\boldsymbol{x}+\lambda\boldsymbol{e}+\bar{\lambda}\bar{\boldsymbol{e}}$, $\boldsymbol{x} \in E_{r,s}$, $\lambda, \bar{\lambda} \in \mathbb{R}$, and we can consider the linear map

$$m : E_{r+1,s+1} \to \mathcal{G}_{r,s}(2), \quad \boldsymbol{x} + \lambda\boldsymbol{e} + \bar{\lambda}\bar{\boldsymbol{e}} \mapsto \boldsymbol{x}I_2 + \lambda\mathfrak{e} + \bar{\lambda}\bar{\mathfrak{e}},$$

where $\mathfrak{e} = \begin{pmatrix} 1 & 0 \\ 0 & -1 \end{pmatrix}$ and $\bar{\mathfrak{e}} = \begin{pmatrix} 0 & 1 \\ -1 & 0 \end{pmatrix}$. Since $\mathfrak{e}^2 = -\bar{\mathfrak{e}}^2 = I_2$, $\mathfrak{e}\bar{\mathfrak{e}} = -\bar{\mathfrak{e}}\mathfrak{e}$, we infer that

$$m(\boldsymbol{x} + \lambda\boldsymbol{e} + \bar{\lambda}\bar{\boldsymbol{e}})^2 = (\boldsymbol{x}^2 + \lambda^2 - \bar{\lambda}^2)I_2 = q(\boldsymbol{x} + \lambda\boldsymbol{e} + \bar{\lambda}\bar{\boldsymbol{e}})I_2.$$

Now the argument proceeds much like in the previous exercise. By the universal property E.3.3, there is a unique algebra homomorphism $\mathcal{G}_{r+1,s+1} \to \mathcal{G}_{r,s}(2)$ that extends m, and this homomorphism has the claimed form. The second part follows from the fact that $(a + b\boldsymbol{e} + \bar{b}\bar{\boldsymbol{e}} + c\boldsymbol{e}\bar{\boldsymbol{e}})\hat{} = \hat{a} - \hat{b}\boldsymbol{e} - \hat{\bar{b}}\bar{\boldsymbol{e}} + \hat{c}\boldsymbol{e}\bar{\boldsymbol{e}}$.

Remark The reverse involution satisfies

$$(a + b\boldsymbol{e} + \bar{b}\bar{\boldsymbol{e}} + c\boldsymbol{e}\bar{\boldsymbol{e}})\tilde{} = \tilde{a} + b^*\boldsymbol{e} + \bar{b}^*\bar{\boldsymbol{e}} - \tilde{c}\boldsymbol{e}\bar{\boldsymbol{e}},$$

where $b^* = \hat{\tilde{b}}$. For example, $\widetilde{b\boldsymbol{e}} = \boldsymbol{e}\tilde{b} = \hat{\tilde{b}}\boldsymbol{e}$, as in this situation $\boldsymbol{e} \cdot b = \bar{\boldsymbol{e}} \cdot b = 0$ for all $b \in \mathcal{G}_{r,s}$ (by 3.3.6). Using the notations above, in place of $x = a + b$ and $z = a - b$ the reverse has $\tilde{a} + b^*$ and $\tilde{a} - b^*$, and $-\tilde{c} + \bar{b}^*$, $-\tilde{c} - \bar{b}^*$ in place of $c + \bar{b}$ and $c - \bar{b}$, all of which do not have a convenient description in terms of x, z, y and w. Consequently, in general there is no a manageable description of the reverse operator in terms of matrices, but note that for $\mathcal{G}_{1,1}$ the reverse is represented by (see Eq. (3.9)) $\widetilde{\begin{pmatrix} x & y \\ w & z \end{pmatrix}} = \begin{pmatrix} x & -w \\ -y & z \end{pmatrix}$.

E.3.7 (The Pauli Representation and the Isomorphism $\mathcal{G}_3 \simeq \mathbb{C}(2)$) Let $\boldsymbol{e}_1, \boldsymbol{e}_2, \boldsymbol{e}_3$ be an orthonormal basis of E_3. An element of $\mathcal{G}_3$ can be uniquely written in the form $\boldsymbol{\xi}_0 + \boldsymbol{\xi}_1\boldsymbol{e}_1 + \boldsymbol{\xi}_2\boldsymbol{e}_2 + \boldsymbol{\xi}_3\boldsymbol{e}_3$, where $\boldsymbol{\xi}_0, \boldsymbol{\xi}_1, \boldsymbol{\xi}_2, \boldsymbol{\xi}_3 \in \mathbf{C}$ (see E.3.2). Show that the $\mathbb{R}$-linear isomorphism $\mathcal{G}_3 \to \mathbb{C}(2)$,

$$\boldsymbol{\xi}_0 + \boldsymbol{\xi}_1\boldsymbol{e}_1 + \boldsymbol{\xi}_2\boldsymbol{e}_2 + \boldsymbol{\xi}_3\boldsymbol{e}_3 \mapsto \xi_0\sigma_0 + \xi_1\sigma_1 + \xi_2\sigma_2 + \xi_3\sigma_3,$$

where $\xi_i \in \mathbb{C}$ denotes the complex number corresponding to $\boldsymbol{\xi}_i$ and

$$\sigma_0 = I_2 = \begin{pmatrix} 1 & 0 \\ 0 & 1 \end{pmatrix}, \ \sigma_1 = \begin{pmatrix} 0 & 1 \\ 1 & 0 \end{pmatrix}, \ \sigma_2 = \begin{pmatrix} 0 & -i \\ i & 0 \end{pmatrix}, \ \sigma_3 = \begin{pmatrix} 1 & 0 \\ 0 & -1 \end{pmatrix}$$

are the so-called *Pauli matrices*, is an algebra isomorphism.

Hint The Pauli matrices satisfy Clifford's relations:

$$\sigma_j^2 = \sigma_0, \quad \sigma_j\sigma_k + \sigma_k\sigma_j = 0 \text{ if } j \neq k.$$

It follows that the linear map $m : E_3 \to \mathbb{C}(2)$,

$$\boldsymbol{x} := \alpha_1\boldsymbol{e}_1 + \alpha_2\boldsymbol{e}_2 + \alpha_3\boldsymbol{e}_3 \mapsto m(\boldsymbol{x}) := \alpha_1\sigma_1 + \alpha_2\sigma_2 + \alpha_3\sigma_3$$

satisfies $m(\boldsymbol{x})^2 = \alpha_1^2 + \alpha_2^2 + \alpha_3^2 = q(\boldsymbol{x})$. Thus there is a unique extension of m to an algebra homomorphism $\mathcal{G}_3 \to \mathbb{C}(2)$, and this homomorphism agrees with the linear isomorphism described above.

E.3.8 ($\mathcal{G}_{r,s}^+$, $r > 0$ or $s > 0$) For any signature (r, s), $\mathcal{G}_{r,s} \simeq \mathcal{G}_{r,s+1}^+$ and $\mathcal{G}_{r,s} \simeq \mathcal{G}_{s+1,r}^+$. As a consequence, $\mathcal{G}_{r,s}^+ \simeq \mathcal{G}_{r,s-1}$ if $s > 0$ and $\mathcal{G}_{r,s}^+ \simeq \mathcal{G}_{s,r-1}$ if $r > 0$. In particular we get that the algebras $\mathcal{G}_{r,s}^+$ and $\mathcal{G}_{s,r}^+$ are isomorphic for any signature (r, s).

Hint We may assume that $E_{r,s} \subset E_{r,s+1}$. Take a unit negative vector $\bar{\boldsymbol{e}}$ in $E_{r,s+1}$ orthogonal to $E_{r,s}$ and consider the linear map $\mu : E_{r,s} \to \mathcal{G}_{r,s+1}, \boldsymbol{x} \mapsto \boldsymbol{x}\bar{\boldsymbol{e}}$. Since $\bar{\boldsymbol{e}}$ anticommutes with $\boldsymbol{x}$, we have $(\bar{\boldsymbol{e}}\boldsymbol{x})^2 = \boldsymbol{x}^2$, and hence there exists a unique algebra homomorphism $\bar{\mu} : \mathcal{G}_{r,s} \to \mathcal{G}_{r,s+1}$ extending μ. Since $\boldsymbol{x}\bar{\boldsymbol{e}}$ is even, the image of $\bar{\mu}$ is contained in $\mathcal{G}_{r,s+1}^+$. In fact, the image of $\bar{\mu}$ is $\mathcal{G}_{r,s+1}^+$ because this algebra also contains, for any $\boldsymbol{x}, \boldsymbol{y} \in E_{r,s}$, $\boldsymbol{x}\boldsymbol{y} = (\bar{\boldsymbol{e}}\boldsymbol{x})(\bar{\boldsymbol{e}}\boldsymbol{y})$. Since $\mathcal{G}_{r,s}$ and $\mathcal{G}_{r,s+1}^+$ have the same dimension, it follows that $\bar{\mu}$ is an algebra isomorphism.

For the second isomorphism, assume $E_{s,r} \subset E_{s+1,r}$, that $\boldsymbol{e}$ is a positive unit vector in $E_{s+1,r}$ orthogonal to $E_{s,r}$ and regard $E_{r,s}$ as $E_{s,r}$ with the metric $\bar{q} = -q$ if the metric of $E_{s,r}$ is q. Now the map $\mu : E_{r,s} \to \mathcal{G}_{s+1,r}, \boldsymbol{x} \mapsto \boldsymbol{x}\boldsymbol{e}$, satisfies

$(\boldsymbol{ex})^2 = -\boldsymbol{x}^2$ and hence there is a unique algebra homomorphism $\bar{\mu} : \mathcal{G}_{r,s} \to \mathcal{G}_{s+1,r}$ extending μ. Now the argument proceeds as in the previous case.

Remark Instead of the products $\boldsymbol{x}\bar{\boldsymbol{e}}$ and $\boldsymbol{xe}$ used to define μ, we could use $\bar{\boldsymbol{e}}\boldsymbol{x}$ and $\boldsymbol{ex}$. The proofs are the same.

E.3.9 (Existence of Folded Geometric Algebras) In this exercise we will show the existence of folded geometric algebras for any special signature (r, s) as a consequence of the properties of the full geometric algebra $\mathcal{G}_{r,s} = (\mathcal{G}, E)$ of signature (r, s). At this stage, the proof of all statements below should be quite straightforward.

Let $I = \boldsymbol{e}_N$ be the pseudoscalar associated with an orthonormal basis of E. Since the signature is special, $n = r + s$ is odd and hence I is central (that is, it commutes with any other element) and $I^2 = 1$. Define $u := \frac{1}{2}(1 + I)$ and $u' := \frac{1}{2}(1 - I)$. Note that changing the orientation of E has the effect of interchanging u and u'.

(1) Check that u and u' satisfy the relations (compare with the elements u and u' introduced for $\mathcal{G}_1$ in E.3.4, p. 58):

$$u + u' = 1, \quad u^2 = u = uI, \quad u'^2 = u' = -Iu', \quad uu' = u'u = 0.$$

(2) Show that $u\mathcal{G}$ and $u'\mathcal{G}$ are at the same ideals of $\mathcal{G}$ and algebras with the induced product, with units u and u', respectively.
(3) Using that $\mathcal{G}^+ \to \mathcal{G}^-$ and $\mathcal{G}^- \to \mathcal{G}^+$, $x \mapsto Ix$, are linear isomorphisms (because I is odd), deduce that $u\mathcal{G} = u\mathcal{G}^+ = u\mathcal{G}^-$ and $u'\mathcal{G} = u'\mathcal{G}^+ = u'\mathcal{G}^-$, and hence that the maps $\mathcal{G}^+ \to u\mathcal{G}$ and $\mathcal{G}^+ \to u'\mathcal{G}$ are algebra isomorphisms.
(4) Prove that the subspace $uE \subseteq u\mathcal{G}$ has dimension n, that $(u\mathcal{G}, uE)$ satisfies the axioms **A0** to **A3**, but that its pseudoscalar is u. Similar statements are valid for the algebra $(u'\mathcal{G}, u'E)$.
(5) Remark however that the isomorphism $\mathcal{G}^+ \simeq u\mathcal{G}$, together with E.3.8, shows that $u\mathcal{G}$ is a full geometric algebra of signature $(r, s-1)$ if $s > 0$ or of signature $(s, r-1)$ if $r > 0$.
(6) The map $\mathcal{G} \to u\mathcal{G} \times u'\mathcal{G}$, $x \mapsto (ux, u'x)$, is an algebra isomorphism. In particular we conclude that $\mathcal{G} \simeq 2\mathcal{G}^+$.

Chapter 4
Orthogonal Geometry with GA

The aim of this chapter is to explore the remarkable way by which the isometry group $\mathrm{O}_q = \mathrm{O}_q(E)$ of an *orthogonal space* (E, q) and its most significant subgroups are accounted for by the formalism of the geometric algebra $(\mathcal{G}, E)$.

Within Klein's classical view of geometry, the study of the groups O_q is what here is denominated *orthogonal geometry*, and the relevance of $(\mathcal{G}, E)$ for its study is the capacity of $\mathcal{G}$ to express both geometric objects and their transformations by elements of O_q.

In the first section, after a short summary of the main notions and notations we need, we provide the description of axial symmetries and reflections by means of the geometric product. The elementary character of these descriptions should facilitate the appreciation of the fitting expressiveness of the formalism. The section ends with a short proof of the celebrated *Cartan–Dieudonné theorem*, a crucial cornerstone for what follows.

The next two sections are devoted to the study of some relevant subgroups of the group $\mathcal{G}^\times$ of invertible elements of $\mathcal{G}$ and their bearing for the study of O_q and some of its subgroups.

In Sect. 4.2, the *versor group* is defined by means of a slight variation of Lipschitz original definition [103, 142], and then it is shown that it is the subgroup of $\mathcal{G}^\times$ generated by the set $E^\times$ of non-null vectors. The *pinor group* is defined as the group of unit versors and then it is seen to be the subgroup generated by unit vectors.

Section 4.3 is devoted to the *spinor group*, which is the group of even pinors, and to the *rotor group*, which is formed by the spinors R such that $R\tilde{R} = 1$ (*rotors*). The section ends with a few remarks and results about the principle of *geometric covariance*.

In the last Sect. 4.4, the formalism is illustrated with the special case of rotations in the n-th dimensional Euclidean space. The section ends with a few remarks about effective computations.

S. Xambó-Descamps, *Real Spinorial Groups*, SpringerBriefs in Mathematics,
https://doi.org/10.1007/978-3-030-00404-0_4

4.1 O and SO

We let $\mathcal{G} = \mathcal{G}_{r,s}$ denote the full geometric algebra of signature (r, s), $n = r + s$. In it, there coexist and interrelate the Grassmann's grading

$$\mathcal{G} = \mathcal{G}^0 \oplus \mathcal{G}^1 \oplus \mathcal{G}^2 \oplus \cdots \oplus \mathcal{G}^n$$

and the products xy (geometric), $x \wedge y$ (outer or exterior), and $x \cdot y$ (inner or interior). The even algebra $\mathcal{G}^+$ consists of multivectors with only even grades and it is a subalgebra for the three products. We also have the involutions $x \mapsto \hat{x}$ and $x \mapsto \tilde{x}$ (parity and reverse involutions), which are, respectively, an automorphism (see 2.1.11, 2.4.4 and Eq. (3.8)) and an antiautomorphism (see 2.1.12, 2.4.5, and Eq. (3.8)) of $\mathcal{G}$.

Let us observe that the multiplicative group of scalars, $\mathbb{R}^\times = \mathbb{R} - \{0\}$, is a subgroup of the group $\mathcal{G}^\times$ of multivectors that are invertible with respect to the geometric product. The latter group also contains the set $E^\times$ of non-null vectors and, more generally, the set of non-null blades (see 3.3.10).

4.1.1 (Extension of an Isometry of E to an Isometry of $\mathcal{G}$) Given $f \in \mathrm{O}_q(E)$, we can apply the universal property of $\mathcal{G}$ (see 3.1.11) to conclude that there is a unique algebra automorphism of $\mathcal{G}$ that extends f and which we will denote by the same symbol f. It is an automorphism of $\mathcal{G}$ in the strong sense that it *preserves the Grassmann's grading* (as f maps the product of k pair-wise orthogonal vectors to a similar product), *the outer and inner products* (as they are defined in terms of the geometric product: see 3.3.2 and 3.3.3), and it *commutes with the involutions* (by the formulas (3.7)). Moreover, *f is an isometry of $\mathcal{G}$* because for any $x \in \mathcal{G}$ we have, using 3.3.9,

$$q(fx) = (\widetilde{f(x)}f(x))_0 = ((f\tilde{x})(fx))_0 = (f(\tilde{x}x))_0 = f((\tilde{x}x)_0) = (\tilde{x}x)_0 = q(x).$$

□

4.1.2 (The Special Orthogonal Groups, SO_q) If I is a non-zero pseudoscalar, then

$$q(I) = q(fI) = q(\det(f)I) = \det(f)^2 q(I),$$

and therefore $\det(f) = \pm 1$. The subset

$$\mathrm{SO}_q(E) = \{f \in \mathrm{O}_q(E) : \det(f) = 1\}$$

is a normal subgroup of $\mathrm{O}_q(E)$, as it is the kernel of the group homomorphism $\det : \mathrm{O}_q \to \{\pm 1\}$. It is called the *special orthogonal group* of q and its elements are said to be *proper isometries* or *rotations*. The group $\mathrm{SO}_q = \mathrm{SO}_q(E)$ is also denoted by $\mathrm{SO}_{r,s}$ (or just SO_n in the case of the Euclidean space of dimension n) if we want to specify the signature.

Axial Symmetries

The *axial symmetry* of E with axis $\langle \boldsymbol{u} \rangle$ ($\boldsymbol{u}$ a non-null vector) is the unique *linear* map $s_{\boldsymbol{u}} : E \to E$ such that $s_{\boldsymbol{u}}(\boldsymbol{u}) = \boldsymbol{u}$ and $s_{\boldsymbol{u}}\boldsymbol{x} = -\boldsymbol{x}$ for all $\boldsymbol{x} \in \boldsymbol{u}^{\perp}$ (for this definition to make sense and work we need that $\boldsymbol{u}$ be non-null, for this insures that $E = \langle \boldsymbol{u} \rangle + \boldsymbol{u}^{\perp}$). The symmetry $s_{\boldsymbol{u}}$ can also be described as the rotation by π about $\boldsymbol{u}$.

If we decompose a vector $\boldsymbol{x}$ as $\boldsymbol{x} = \lambda\boldsymbol{u} + \boldsymbol{x}'$, with $\boldsymbol{x}' \in \boldsymbol{u}^{\perp}$, then $s_{\boldsymbol{u}}\boldsymbol{x} = \lambda\boldsymbol{u} - \boldsymbol{x}'$ and

$$q(s_{\boldsymbol{u}}\boldsymbol{x}) = q(\lambda\boldsymbol{u}) + q(-\boldsymbol{x}') = q(\lambda\boldsymbol{u}) + q(\boldsymbol{x}') = q(\boldsymbol{x}),$$

which shows that $s_{\boldsymbol{u}}$ is an isometry.

Since $\det(s_{\boldsymbol{u}}) = (-1)^{n-1}$, we see that $s_{\boldsymbol{u}}$ is a proper isometry if and only if n is odd.

4.1.3 (GA Expression of an Axial Symmetry) *For any vector $\boldsymbol{x}$, $s_{\boldsymbol{u}}\boldsymbol{x} = \boldsymbol{u}\boldsymbol{x}\boldsymbol{u}^{-1}$.*

Proof The map $\boldsymbol{x} \mapsto \boldsymbol{u}\boldsymbol{x}\boldsymbol{u}^{-1}$ is linear in $\boldsymbol{x}$ and clearly leaves $\boldsymbol{u}$ invariant. Now if $\boldsymbol{x} \in \boldsymbol{u}^{\perp}$, then $\boldsymbol{u}$ and $\boldsymbol{x}$ anticommute and hence $\boldsymbol{u}\boldsymbol{x}\boldsymbol{u}^{-1} = -\boldsymbol{x}\boldsymbol{u}\boldsymbol{u}^{-1} = -\boldsymbol{x}$. □

Reflections

Given a non-null vector $\boldsymbol{u}$, the *reflection of E in the direction $\boldsymbol{u}$*, or *across the hyperplane $\boldsymbol{u}^{\perp}$*, is the linear map $m_{\boldsymbol{u}} : E \to E$ such that $m_{\boldsymbol{u}}(\boldsymbol{u}) = -\boldsymbol{u}$ and $m_{\boldsymbol{u}}\boldsymbol{x} = \boldsymbol{x}$ for all $\boldsymbol{x} \in \boldsymbol{u}^{\perp}$. In other words, $m_{\boldsymbol{u}} = -s_{\boldsymbol{u}}$, and hence $m_{\boldsymbol{u}}$ is also an isometry. This isometry is always improper, for $\det(m_{\boldsymbol{u}}) = (-1)^n \det(s_{\boldsymbol{u}}) = -1$.

4.1.4 (GA Expression of a Reflection) *For any vector $\boldsymbol{x}$, $m_{\boldsymbol{u}}\boldsymbol{x} = -\boldsymbol{u}\boldsymbol{x}\boldsymbol{u}^{-1}$.* □

4.1.5 (Remark) For any non-null vector $\boldsymbol{u}$ and any non-zero scalar λ, $\boldsymbol{u}$ and $\lambda\boldsymbol{u}$ plainly define the same axial symmetry, and hence also the same reflection. Taking $\lambda = 1/|\boldsymbol{u}|$, in which case $q(\lambda\boldsymbol{u}) = \boldsymbol{u}^2/|\boldsymbol{u}|^2 = \epsilon_{\boldsymbol{u}}$ (the signature of $\boldsymbol{u}$), we see that to describe an axial symmetry $s_{\boldsymbol{u}}$, or the reflection $m_{\boldsymbol{u}}$, we can always assume that $\boldsymbol{u}$ is a unit vector.

4.1.6 (Alternative Form of $m_{\boldsymbol{u}}$) For effective computations, it is important to observe that if $\boldsymbol{u}$ is a unit vector, then

$$m_{\boldsymbol{u}}(\boldsymbol{x}) = \boldsymbol{x} - 2\epsilon_{\boldsymbol{u}}\boldsymbol{u}(\boldsymbol{u} \cdot \boldsymbol{x}).$$

Indeed, the right-hand side is linear in $\boldsymbol{x}$, leaves $\boldsymbol{x}$ fixed if $\boldsymbol{x} \in \boldsymbol{u}^{\perp}$, and it maps $\boldsymbol{u}$ to $\boldsymbol{u} - 2\epsilon_{\boldsymbol{u}}\boldsymbol{u}(\boldsymbol{u} \cdot \boldsymbol{u}) = -\boldsymbol{u}$, as $\boldsymbol{u} \cdot \boldsymbol{u} = \boldsymbol{u}^2 = \epsilon_{\boldsymbol{u}}$. □

The Cartan–Dieudonné Theorem

We need to establish that any q-isometry can be expressed as a product of reflections. This is a weak form of the Cartan–Dieudonné theorem, which says, in its strong version, that no more than n reflections are needed. Our approach to this theorem will be to provide a proof of the strong form, for all the steps in a proof of the weak form are actually valid for the strong form, except that the latter requires an extra step. Our proof is simpler than the original one (cf. [34]), or than later variations, as for example [61]. Note also that some authors, as for example [52], include a proof of the weak form, but give a reference, usually [34], for a proof of the strong form.

4.1.7 (The Cartan–Dieudonné Theorem) *Any* $f \in \mathrm{O}_{r,s}$, $f \neq \mathrm{Id}$, *is a composition of at most n reflections.*

Proof We divide the proof into four steps.

(0) We may assume that $n \geqslant 2$, as for $n = r + s = 1$ we have $f = -\mathrm{Id}$, which in this case is a reflection. Thus the theorem is valid for $n = 1$. Now we can argue by induction on n. We distinguish three cases.

(1) If there is a non-null vector $\boldsymbol{u}$ such that $f(\boldsymbol{u}) = \boldsymbol{u}$, then f induces an isometry f' of $\boldsymbol{u}^{\perp}$, which by induction is a composition of at most $n-1$ reflections. It follows that f itself is a composition of at most $n-1$ reflections, because each reflection m' of $\boldsymbol{u}^{\perp}$ extends to a unique reflection m of E such that $m(\boldsymbol{u}) = \boldsymbol{u}$.

(2) Thus we may assume that $f(\boldsymbol{u}) \neq \boldsymbol{u}$, or $d(\boldsymbol{u}) := f(\boldsymbol{u}) - \boldsymbol{u} \neq 0$, for all non-null vectors $\boldsymbol{u}$. If $\boldsymbol{v} := d(\boldsymbol{u})$ is non-null for some non-null $\boldsymbol{u}$, then $m_{\boldsymbol{v}}(f\boldsymbol{u}) = \boldsymbol{u}$, because on one hand

$$-f(\boldsymbol{u}) + \boldsymbol{u} = -\boldsymbol{v} = m_{\boldsymbol{v}}(\boldsymbol{v}) = m_{\boldsymbol{v}}(f\boldsymbol{u}) - m_{\boldsymbol{v}}(\boldsymbol{u})$$

and on the other, since $\boldsymbol{v}' := f(\boldsymbol{u}) + \boldsymbol{u}$ is orthogonal to $\boldsymbol{v}$,

$$f(\boldsymbol{u}) + \boldsymbol{u} = \boldsymbol{v}' = m_{\boldsymbol{v}}(\boldsymbol{v}') = m_{\boldsymbol{v}}(f\boldsymbol{u}) + m_{\boldsymbol{v}}(\boldsymbol{u}),$$

and the relation follows by a simple addition. Now by step (1) we have that $g = m_{\boldsymbol{v}} f$ is the composition of at most $n-1$ reflections and hence $f = m_{\boldsymbol{v}} g$ is the composition of at most n reflections.

(3) It remains to analyze the possibility that $d(\boldsymbol{u})$ *is null and non-zero for any non-null* $\boldsymbol{u}$. In this situation we claim that $d(\boldsymbol{x})$ is null for all vectors $\boldsymbol{x}$. Indeed, we only need to see that $d(\boldsymbol{x})$ is also null when $\boldsymbol{x}$ is null. Let $\boldsymbol{u}$ be a non-null vector and $\lambda \in \mathbb{R}$, $\lambda \neq 0, -2(\boldsymbol{x} \cdot \boldsymbol{u})/\boldsymbol{u}^2$. Then $\boldsymbol{x} + \lambda \boldsymbol{u}$ is non-null, because $(\boldsymbol{x} + \lambda \boldsymbol{u})^2 = \lambda(2\boldsymbol{x} \cdot \boldsymbol{u} + \lambda \boldsymbol{u}^2) \neq 0$. So $d(\boldsymbol{x} + \lambda \boldsymbol{u}) = d(\boldsymbol{x}) + \lambda d(\boldsymbol{u})$ is null (and non-zero). But

$$0 = (d(\boldsymbol{x}) + \lambda d(\boldsymbol{u}))^2 = d(\boldsymbol{x})^2 + 2\lambda d(\boldsymbol{x}) \cdot d(\boldsymbol{u})$$

cannot happen for more than one value of λ unless $d(\boldsymbol{x})^2 = 0$ (and $d(\boldsymbol{x}) \cdot d(\boldsymbol{u}) = 0$ as well), and this proves the claim.

Thus all vectors of $D := d(E)$ are null, and hence we have, by the polarization formula, $d(\boldsymbol{x}) \cdot d(\boldsymbol{y}) = 0$ for all $\boldsymbol{x}, \boldsymbol{y} \in E$. Using the identity $\boldsymbol{x} + d(\boldsymbol{x}) = f(\boldsymbol{x})$, which is valid for any vector $\boldsymbol{x}$, we get that

$$\boldsymbol{x} \cdot \boldsymbol{y} = f(\boldsymbol{x}) \cdot f(\boldsymbol{y}) = (\boldsymbol{x} + d(\boldsymbol{x})) \cdot (\boldsymbol{y} + d(\boldsymbol{y})) = \boldsymbol{x} \cdot \boldsymbol{y} + \boldsymbol{x} \cdot d(\boldsymbol{y}) + d(\boldsymbol{x}) \cdot \boldsymbol{y},$$

and thereby $\boldsymbol{x} \cdot d(\boldsymbol{y}) + \boldsymbol{y} \cdot d(\boldsymbol{x}) = 0$ for all $\boldsymbol{x}, \boldsymbol{y} \in E$. If we apply this identity to $\boldsymbol{x} \in D^{\perp}$, we get $\boldsymbol{y} \cdot d(\boldsymbol{x}) = 0$ for all $\boldsymbol{y}$, and so $d(\boldsymbol{x}) = 0$, which means that $f(\boldsymbol{x}) = \boldsymbol{x}$, for any $\boldsymbol{x} \in D^{\perp}$. By the hypothesis of the current case, all vectors in $D^{\perp}$ must be isotropic, and this implies that $D = D^{\perp}$. Indeed, we have $D \subseteq D^{\perp}$, because D is totally isotropic, and $D^{\perp} \subseteq D^{\perp\perp} = D$, where the first inclusion is justified because $D^{\perp}$ is totally isotropic and the second because $D \subseteq D^{\perp\perp}$ and both spaces have the same dimension (see 1.3.3). Note that $n = 2r$, where $r = \dim(D) = \dim(D^{\perp})$.

Now f induces the identity on $D^{\perp} = D$ and the identity on E/D, because

$$f(\boldsymbol{x} + D) = f(\boldsymbol{x}) + D = \boldsymbol{x} + d(\boldsymbol{x}) + D = \boldsymbol{x} + D.$$

This implies that $f \in SO_{r,s}$ (see E.4.1, p. 75). For any reflection m, $g = mf \notin \mathrm{SO}_{r,s}$ cannot fall on case (3), and so it is a product of at most $n = 2r$ reflections by the previous cases, and since the number of reflections is odd, it cannot be $2r$, and so it is at most $n - 1$. It follows that $f = mg$ is the product of at most n reflections.

□

4.2 Versors and Pinors

A *versor* of $E_{r,s}$ is a multivector $v \in \mathcal{G}^+ \sqcup \mathcal{G}^-$ (so v is an even or odd multivector), which is *invertible* (so $v \in \mathcal{G}^{\times}$) and satisfies $vEv^{-1} = E$ (equivalently, $vEv^{-1} \subseteq E$).

The set of versors will be denoted by $\mathcal{V} = \mathcal{V}_{r,s}$. It is clear that $\mathbb{R}^{\times} \subseteq \mathcal{V}$. It is also clear that $E^{\times} \subseteq \mathcal{V}$, where $E^{\times}$ is the set on non-isotropic (or non-null) vectors, because we know that if $\boldsymbol{v} \in E^{\times}$ then $\boldsymbol{v} \in \mathcal{G}^-$ and $\boldsymbol{v}\boldsymbol{x}\boldsymbol{v}^{-1} = s_{\boldsymbol{v}}(\boldsymbol{x}) \in E$.

4.2.1 (Remark) If in the definition of versor the condition $v \in \mathcal{G}^+ \sqcup \mathcal{G}^-$ is dropped, the seemingly more general notion may be called a *Lipschitz versor* ([103]; cf. [142]). The reason for our choice is that it favors a substantially simpler presentation and that in fact it turns out to be not restrictive at all (see E.4.2, p. 75). The simplification stems from the fact that $\hat{v} = \pm v$ for any versor v, and hence we have $\underline{v}(E) = E$, where $\underline{v} : E \to E$ is defined by

$$\underline{v}(\boldsymbol{x}) = \hat{v}\boldsymbol{x}v^{-1} \text{ for all } \boldsymbol{x} \in E. \tag{4.1}$$

Now we are ready to state and prove the fundamental result of this section, which in particular shows that $\mathcal{V}$ is a group (the *versor group* or *Lipschitz versor group*) and that its elements coincide with the multivectors that are the product of non-isotropic vectors.

4.2.2 (The Lipschitz's Group and Its Twisted Representation)

(1) *The set $\mathcal{V}$ is a subgroup of $\mathcal{G}^{\times}$. In particular it contains any product of non-isotropic vectors.*
(2) *For any versor v, $\underline{v} \in \mathrm{O}_{r,s}$.*

(3) $\tilde{\rho} : \mathcal{V} \to \mathrm{O}_{r,s}$, $v \mapsto \underline{v}$, *is an onto group homomorphism* (called the *twisted adjoint representation*).
(4) *The kernel of* $\tilde{\rho}$ *is* $\mathbb{R}^\times$.
(5) $\mathcal{V}$ *is generated by* $E^\times$. *Hence any versor is the product of non-isotropic vectors.*

Proof

(1) Indeed: $1 \in \mathcal{V}$, and the product uv of $u, v \in \mathcal{V}$ lies in $\mathcal{G}^+ \sqcup \mathcal{G}^-$ (this set is obviously closed under the geometric product) and

$$(uv)E(uv)^{-1} = uvEv^{-1}u^{-1} = u(vEv^{-1})u^{-1} = uEu^{-1} = E.$$

Note also that if $vEv^{-1} = E$, then $v^{-1}Ev = E$, which means that $v^{-1} \in \mathcal{V}$ if $v \in \mathcal{V}$.
(2) The map $\underline{v}$ is plainly linear, and since $\hat{v} = \pm v$ we have

$$q(\underline{v}\boldsymbol{x}) = (\underline{v}\boldsymbol{x})^2 = \hat{v}\boldsymbol{x}v^{-1}\hat{v}\boldsymbol{x}v^{-1} = v\boldsymbol{x}v^{-1}v\boldsymbol{x}v^{-1} = \boldsymbol{x}^2.$$

(3) The map $\tilde{\rho}$ is a homomorphism, because if $u, v \in \mathcal{V}$, then $\underline{uv} = \underline{u}\,\underline{v}$:

$$\underline{uv}(\boldsymbol{x}) = \widehat{uv}\,\boldsymbol{x}(uv)^{-1} = \hat{u}\hat{v}\boldsymbol{x}v^{-1}u^{-1} = \underline{u}(\underline{v}\boldsymbol{x}) = (\underline{u}\,\underline{v})(\boldsymbol{x}).$$

Now if $\boldsymbol{v}$ is a non-isotropic vector, we have, using 4.1.4, that $\tilde{\rho}(\boldsymbol{v}) = m_{\boldsymbol{v}}$, for

$$\underline{\boldsymbol{v}}(\boldsymbol{x}) = \hat{\boldsymbol{v}}\boldsymbol{x}\boldsymbol{v}^{-1} = -\boldsymbol{v}\boldsymbol{x}\boldsymbol{v}^{-1} = m_{\boldsymbol{v}}(\boldsymbol{x}).$$

Thus the image of $\tilde{\rho}$ contains the isometries that can be expressed as a product of reflections, which happens for all isometries by the Cartan–Dieudonné Theorem 4.1.7. Therefore $\tilde{\rho}$ is onto.
(4) Indeed, if $v \in \ker(\tilde{\rho})$, $\hat{v}\boldsymbol{x} = \boldsymbol{x}v$ for all $\boldsymbol{x} \in E$, which by Riesz's second formula is equivalent to say that $\boldsymbol{x} \cdot v = 0$ for all $\boldsymbol{x} \in E$ (see 3.3.6). Now 2.4.11 tells us that v must be a scalar.
(5) If $v \in \mathcal{V}$, there exists $\boldsymbol{u}_1, \ldots, \boldsymbol{u}_k \in E^\times$ such that $\tilde{\rho}(v) = \tilde{\rho}(u)$, where $u = \boldsymbol{u}_1 \cdots \boldsymbol{u}_k$. This means that $\lambda = vu^{-1} \in \ker(\tilde{\rho}) = \mathbb{R}^\times$ and consequently $v = \lambda u$, which is a product of non-isotropic vectors. □

4.2.3 (Example: $\underline{I}$) Let $I \in \mathcal{G}^n$ be a non-zero pseudoscalar. Then $I \in \mathcal{V}$ and $\underline{I} = -\mathrm{Id}$. Indeed, I is clearly a versor, and for any vector $\boldsymbol{x}$ we have

$$\underline{I}(\boldsymbol{x}) = \hat{I}\boldsymbol{x}I^{-1} = (-1)^n I\boldsymbol{x}I^{-1} \stackrel{*}{=} (-1)^n(-1)^{n+1}\boldsymbol{x}I\,I^{-1} = -\boldsymbol{x},$$

where in the step $\stackrel{*}{=}$ we have set $I\boldsymbol{x} = (-1)^{n+1}\boldsymbol{x}I$, which is justified by 3.4.1, statement 3. For an alternative proof, see E.4.3, p. 75.

4.2.4 (The Group of Even Versors) *Consider the subgroup* $\mathcal{V}^+ = \mathcal{V}^+_{r,s}$ *of* $\mathcal{V}_{r,s}$ *formed by the even versors. For any* $v \in \mathcal{V}^+$, $\underline{v}$ *is the product of an even number* *of reflections and hence* $\underline{v} \in \mathrm{SO}_{r,s}$. *The map* $\tilde{\rho} : \mathcal{V}^+_{r,s} \to \mathrm{SO}_{r,s}$ *is an onto group homomorphism* (again by the Cartan–Dieudonné theorem), *and its kernel is* $\mathbb{R}^\times$. □

Given a versor v, the alternative form for the metric of $\mathcal{G}$ (see 3.3.9), and the fact that $v\tilde{v}$ is a scalar, tells us that

$$q(v) = v\tilde{v} = \epsilon(v)|v|^2, \tag{4.2}$$

where $\epsilon(v)$ is the sign of $q(v)$ and $|v|$ is the *magnitude* of v. Note that Eq. (4.2) is equivalent to

$$\tilde{v} = \epsilon(v)|v|^2 v^{-1} \text{ or } \tilde{v}^{-1} = \epsilon(v)|v|^{-2} v. \tag{4.3}$$

The group Pin $=$ $\mathrm{Pin}_{r,s}$ is defined as the subgroup of $\mathcal{V}$ formed by the *unit versors*, that is, the versors v such that $q(v) = v\tilde{v} = \pm 1$. The elements of Pin are called *pinors*.

4.2.5 *The group* Pin *coincides with the subgroup of* $\mathcal{V}$ *formed by the products of unit vectors.*

Proof Since the product of unit vectors is clearly a pinor, all is reduced to show the converse.

Let $v = \boldsymbol{u}_1 \cdots \boldsymbol{u}_k$ be a pinor, $\boldsymbol{u}_1, \ldots, \boldsymbol{u}_k$ non-null vectors. Since

$$q(v) = v\tilde{v} = \epsilon,\ \epsilon = \pm 1,$$

it turns out that $q(\boldsymbol{u}_1) \cdots q(\boldsymbol{u}_k) = \epsilon$.

Let $\lambda_j = |\boldsymbol{u}_j| > 0$ and $\epsilon_j = \epsilon_{\boldsymbol{u}_j}$, so that $q(\boldsymbol{u}_j) = \boldsymbol{u}_j^2 = \epsilon_j \lambda_j^2$. Then

$$\epsilon = \epsilon_1 \cdots \epsilon_k \lambda_1^2 \cdots \lambda_k^2,$$

which implies that $\lambda_1 \cdots \lambda_k = 1$. Consequently, $v = \boldsymbol{u}'_1 \cdots \boldsymbol{u}'_k$, with $\boldsymbol{u}'_j = \boldsymbol{u}_j/\lambda_j$, and $\boldsymbol{u}'_j$ is a unit vector: $q(\boldsymbol{u}'_j) = q(\boldsymbol{u}_j)/\lambda_j^2 = \epsilon_j$. □

4.2.6 *The homomorphism* $\tilde{\rho} : \mathrm{Pin}_{r,s} \to \mathrm{O}_{r,s}$, $v \mapsto \underline{v}$, *is onto and its kernel is* $\{\pm 1\}$.

Proof The first part is a consequence of the Cartan–Dieudonné theorem, because any reflection has the form $m_{\boldsymbol{u}}$ with $\boldsymbol{u}$ a unit vector. The kernel is formed by the scalars λ such that $q(\lambda) = \lambda^2 = 1$, so $\lambda = \pm 1$. □

4.3 Spinors and Rotors

The group Spin $=$ $\mathrm{Spin}_{r,s} = \mathrm{Pin}^+_{r,s}$ is the subgroup of even elements of $\mathrm{Pin}_{r,s}$, that is, pinors that are the product of an even number of unit vectors. The elements of Spin are called *spinors*.

4.3.1 $\tilde{\rho} : \mathrm{Spin}_{r,s} \to \mathrm{SO}_{r,s}$ *is an onto homomorphism and its kernel is* $\{\pm 1\}$. □

Since spinors R are unit pinors, in general we have $R\tilde{R} = \pm 1$. If $R\tilde{R} = 1$, we say that the spinor is a *rotor*. The set of rotors will be denoted by $\mathcal{R}_{r,s}$. In the Euclidean and anti-Euclidean cases ($E_n = E_{n,0}$ and $\bar{E}_n = E_{0,n}$), all spinors are rotors, but otherwise this is not the case: if $r, s > 0$ and $v = \boldsymbol{u}\bar{\boldsymbol{u}}$ with $\boldsymbol{u}^2 = 1$ and $\bar{\boldsymbol{u}}^2 = -1$, then v is a spinor, but not a rotor, for $v\tilde{v} = -1$.

4.3.2 *The set of rotors* $\mathcal{R}_{r,s}$ *is a normal subgroup of* $\mathrm{Spin}_{r,s}$.

Proof Consider the map $q : \mathrm{Spin}_{r,s} \to \{\pm 1\}$, $S \mapsto S\tilde{S} = q(S)$. This map is a group homomorphism, as

$$q(ST) = (ST)(ST)^{\sim} = ST\tilde{T}\tilde{S} = (S\tilde{S})(T\tilde{T}) = q(S)q(T)$$

if S and T are spinors, and its kernel is $\mathcal{R}_{r,s}$. □

The image of $\mathcal{R}_{r,s}$ in $\mathrm{SO}_{r,s}$ by $\tilde{\rho}$ will be denoted by $\mathrm{SO}^0_{r,s}$. Thus we have an onto group homomorphism

$$\tilde{\rho} : \mathcal{R}_{r,s} \to \mathrm{SO}^0_{r,s} \tag{4.4}$$

whose kernel is $\{\pm 1\}$. In the Euclidean and anti-Euclidean cases, $\mathrm{SO}^0_n = \mathrm{SO}_n$, but for $r, s > 0$ we have seen that $\mathrm{SO}^0_{r,s}$ is a proper subgroup of $\mathrm{SO}_{r,s}$.

Primacy of the Rotor Group

In the two statements that follow we show that the rotor group $\mathcal{R}$ determines the structure of the spinor group $\mathcal{S} = \mathrm{Spin}$ and of the pinor group $\mathcal{P} = \mathrm{Pin}$.

4.3.3 *If* $(r, s) = (n, 0)$ (Euclidean case) *or* $(r, s) = (0, n)$ (anti-Euclidean case), *then* $\mathcal{S} = \mathcal{R}$ *and* $\mathcal{P} = \mathcal{R} \sqcup \boldsymbol{u}\mathcal{R}$ *for any given unit vector* $\boldsymbol{u}$. *As a consequence,* $\mathrm{O}_n = \mathrm{SO}_n \sqcup m_{\boldsymbol{u}}\mathrm{SO}_n$.

Proof Let $v = \boldsymbol{u}_1 \cdots \boldsymbol{u}_k \in \mathcal{P}$, where the $\boldsymbol{u}_j$ are unit vectors. If k is even, then $v \in \mathcal{S}$ and $v\tilde{v} = \boldsymbol{u}_1^2 \cdots \boldsymbol{u}_k^2 = 1$ (in the anti-Euclidean case, the product is $(-1)^k = 1$ because k is even). This shows that $v \in \mathcal{R}$. If k is odd, then $v = \boldsymbol{u}(\epsilon_{\boldsymbol{u}}\boldsymbol{u}v)$ and $\epsilon_{\boldsymbol{u}}\boldsymbol{u}v \in \mathcal{R}$. □

4.3.4 *If* $r, s \geqslant 1$, *fix unit vectors* $\boldsymbol{u}$ *and* $\bar{\boldsymbol{u}}$ *such that* $\boldsymbol{u}^2 = 1$ *and* $\bar{\boldsymbol{u}}^2 = -1$. *Then we have:* $\mathcal{S} = \mathcal{R} \sqcup \boldsymbol{u}\bar{\boldsymbol{u}}\mathcal{R}$ *and* $\mathcal{P} = \mathcal{R} \sqcup \boldsymbol{u}\mathcal{R} \sqcup \bar{\boldsymbol{u}}\mathcal{R} \sqcup \boldsymbol{u}\bar{\boldsymbol{u}}\mathcal{R} = \mathcal{S} \sqcup \boldsymbol{u}\mathcal{S}$. *As a consequence,* $\mathrm{SO}_{r,s} = \mathrm{SO}^0_{r,s} \sqcup \rho(\boldsymbol{u}\bar{\boldsymbol{u}})\mathrm{SO}^0_{r,s}$ *and* $\mathrm{O}_{r,s} = \mathrm{SO}_{r,s} \sqcup m_{\boldsymbol{u}}\mathrm{SO}_{r,s}$.

Proof It is clear that $\mathcal{S} = \mathcal{R} \sqcup \mathcal{R}'$, where $v \in \mathcal{R}$ or $v \in \mathcal{R}'$ according to whether $v\tilde{v} = 1$ or $v\tilde{v} = -1$. Now it is easy to check that

$$w \in \mathcal{R}' \Leftrightarrow v = \boldsymbol{u}\bar{\boldsymbol{u}}w \in \mathcal{R}$$

(use that $(\boldsymbol{u}\bar{\boldsymbol{u}})(\boldsymbol{u}\bar{\boldsymbol{u}})^{\sim} = -1$). In a similar way we see that $\boldsymbol{u}\mathcal{R}$ and $\bar{\boldsymbol{u}}\mathcal{R}$ are the odd pinors v such that $v\tilde{v} = +1$ and $v\tilde{v} = -1$, respectively. □

Geometric Covariance

Given an element $v \in \mathcal{G}^{\pm}$, in particular if v is a versor, the *parity* of v is defined to be ± 1, and will be denoted by $\pi(v)$. In other words, $\pi(v) = 1$ if v is even and $\pi(v) = -1$ if v is odd. Notice that for such elements we have $\hat{v} = \pi(v)v$. The map $\pi : \mathcal{V} \to \{\pm 1\}$ is clearly onto and it is easily checked that it is a group homomorphism. Its kernel is the group $\mathcal{V}^+$ of even versors.

4.3.5 (Geometric Covariance) *Let $v \in \mathcal{V}$ and $x, y \in \mathcal{G}$. Then the map $\underline{v} : \mathcal{G} \to \mathcal{G}$ is a linear graded automorphism that has the following additional properties:*

(1) $\underline{v}(\hat{x}) = \widehat{\underline{v}(x)}$ *and* $\underline{v}(\tilde{x}) = \widetilde{\underline{v}(x)}$.
(2) $\underline{v}(xy) = \pi(v)\,\underline{v}(x)\,\underline{v}(y)$.
(3) $\underline{v}(x \wedge y) = \pi(v)\,\underline{v}(x) \wedge \underline{v}(y)$ *and* $\underline{v}(x \cdot y) = \pi(v)\,\underline{v}(x) \cdot \underline{v}(y)$.
(4) *If v is even, then $\underline{v}$ is an automorphism of the geometric algebra, by which we mean that it is a linear automorphism that preserves the grading, that is an automorphism of the geometric, outer and inner products, and that commutes with the parity and reverse involutions.*

Proof

(1) These relations are an immediate consequence of the preservation of grades by $\underline{v}$ and the formulas (3.7) for the parity and reverse involutions.
(2) Indeed,

$$\underline{v}(xy) = \pi(v)v(xy)v^{-1} = \pi(v)\; vxv^{-1}vyv^{-1} = \pi(v)\;\pi(v)vxv^{-1}\pi(v)vyv^{-1}$$
$$= \pi(v)\underline{v}(x)\underline{v}(y).$$

(3) These relations follow from (2) and the expressions of the outer and inner products in terms of the geometric product (see 3.3.8).
(4) In this case, $\pi(v) = 1$ and the claim is an immediate consequence of the preceding statements. □

We end the section with three corollaries of geometric covariance.

4.3.6 (Example: Reflection of Blades) Let $X = \boldsymbol{x}_1 \wedge \cdots \wedge \boldsymbol{x}_k$ be a k-blade and $\boldsymbol{u}$ a unit vector. The reflection of X in the direction $\boldsymbol{u}$ is the blade $m_{\boldsymbol{u}}(\boldsymbol{x}_1) \wedge \cdots \wedge m_{\boldsymbol{u}}(\boldsymbol{x}_k)$, which by geometric covariance is equal to $m_{\boldsymbol{u}}(X) = -\boldsymbol{u}X\boldsymbol{u}$.

Let us also observe that the alternative expression $m_{\boldsymbol{u}}(\boldsymbol{x}) = \boldsymbol{x} - 2\epsilon_{\boldsymbol{u}}\boldsymbol{u}(\boldsymbol{u} \cdot \boldsymbol{x})$ for the reflection of a vector $\boldsymbol{x}$ in the direction $\boldsymbol{u}$ (see 4.1.6) is also valid for blades in the following guise:

$$m_{\boldsymbol{u}}(X) = X - 2\epsilon_{\boldsymbol{u}}\boldsymbol{u} \wedge (\boldsymbol{u} \cdot X). \tag{4.5}$$

Proof If we apply the alternative form of a reflection for the vectors $\boldsymbol{x}_j$, we get

$$m_{\boldsymbol{u}}(X) = (\boldsymbol{x}_1 - 2\epsilon_{\boldsymbol{u}}\boldsymbol{u}(\boldsymbol{u} \cdot \boldsymbol{x}_1)) \wedge \cdots \wedge (\boldsymbol{x}_k - 2\epsilon_{\boldsymbol{u}}\boldsymbol{u}(\boldsymbol{u} \cdot \boldsymbol{x}_k)).$$

Since $\boldsymbol{u} \wedge \boldsymbol{u} = 0$, in the expansion of this product it suffices to choose the second summand of the factors at most once. Setting $X_j = \boldsymbol{x}_1 \wedge \cdots \wedge \boldsymbol{x}_{j-1} \wedge \boldsymbol{x}_{j+1} \wedge \cdots \wedge \boldsymbol{x}_k$, we get

$$m_{\boldsymbol{u}}(X) = X - 2\epsilon_{\boldsymbol{u}} \boldsymbol{u} \wedge \textstyle\sum_j (-1)^{j-1} (\boldsymbol{u} \cdot \boldsymbol{x}_j) X_j.$$

Since $\sum_j (-1)^{j-1} (\boldsymbol{u} \cdot \boldsymbol{x}_j) X_j = \boldsymbol{u} \cdot X$, the proof is complete. □

The formula (4.5) turns out to be an important tool for speeding up the computation of the reflection of objects that are representable as blades and consequently, thanks to the Cartan–Dieudonné theorem, the computation of transformations of those objects by arbitrary isometries.

4.3.7 *For any versor v, the map $\underline{v} : \mathcal{G} \to \mathcal{G}$ is an isometry. It follows that if u is a pinor (spinor), then $\underline{v}(u)$ is a pinor (spinor) of the same signature as u. In particular, $\underline{v}(u)$ is a rotor when u is a rotor.*

Proof We show that $q(\underline{v}(u)) = q(u)$, for any multivector u, by a straightforward computation (in the last step we use that $\underline{v}(1) = \pi(v)$):

$$\begin{aligned} q(\underline{v}(u)) &= (\underline{v}(u)\widetilde{\underline{v}(u)})_0 = (\underline{v}(u)\underline{v}(\tilde{u}))_0 \\ &= (\pi(v)\underline{v}(u\tilde{u}))_0 = (\pi(v)\, \underline{v}(q(u)))_0 = q(u). \end{aligned}$$

□

4.3.8 *Le u and v be versors and write $w = \underline{v}(u)$. Then we have that $\underline{w} = \underline{v}\, \underline{u}\, \underline{v}^{-1}$.*

Proof For any vector $\boldsymbol{x}$ we have:

$$\begin{aligned} \underline{\underline{v}(u)}(\boldsymbol{x}) &= \pi(\underline{v}(u))\; \underline{v}(u)\boldsymbol{x}(\underline{v}(u))^{-1} = \pi(u)\,\pi(v)\, vuv^{-1}\boldsymbol{x}\,(\pi(v)\, vuv^{-1})^{-1} \\ &= \pi(u)\; vuv^{-1}\boldsymbol{x}\; vu^{-1}v^{-1} = \pi(u)\,\pi(v)\,\underline{v}(uv^{-1}\boldsymbol{x}\; vu^{-1}) \\ &= \pi(v)\,\underline{v}(\underline{u}(v^{-1}\boldsymbol{x}\; v)) = \underline{v}(\underline{u}(\underline{v^{-1}}(\boldsymbol{x}))) = \underline{v}(\underline{u}(\underline{v}^{-1}(\boldsymbol{x}))). \end{aligned}$$

□

4.4 Rotations in the Euclidean Space

In this section we apply what we have learned so far to provide a fresh approach to the treatment of rotations in the Euclidean space. One motivation for this section stems from the following quotation[1]

[1] R. Feynman, *Lecture Notes in Physics*, Volume I, Section 22-6.

The most remarkable formula in mathematics:

$$e^{i\theta} = \cos\theta + i\sin\theta.$$

This is our jewel.

So, what has GA to say about this jewel?
Let $\boldsymbol{u}$ and $\boldsymbol{v}$ be linearly independent unit vectors of the Euclidean space E_n and consider the rotor $R = \boldsymbol{vu}$ (see Fig. 4.1). This rotor produces the rotation $\underline{R}$,

$$\underline{R}(\boldsymbol{x}) = R\boldsymbol{x}R^{-1} = R\boldsymbol{x}\tilde{R}.$$

What is the axis of this rotation? What is its rotation angle?

Let $\theta \in (0, \pi)$ be the Euclidean angle between $\boldsymbol{u}$ and $\boldsymbol{v}$:

$$\cos\theta = \boldsymbol{u} \cdot \boldsymbol{v}.$$

Let $\boldsymbol{i}$ be the unit area of the oriented plane $P = \langle \boldsymbol{u}, \boldsymbol{v} \rangle$, that is, $\boldsymbol{i} = \boldsymbol{u}_1\boldsymbol{u}_2$ with $\boldsymbol{u}_1, \boldsymbol{u}_2$ a positive orthonormal basis of P. Then $\boldsymbol{i}^2 = -1$. Moreover, $\boldsymbol{x} \mapsto \boldsymbol{xi}$ is the counterclockwise rotation in P of amplitude $\pi/2$, for $\boldsymbol{u}_1\boldsymbol{i} = \boldsymbol{u}_2$ and $\boldsymbol{u}_2\boldsymbol{i} = -\boldsymbol{u}_1$. In particular $\boldsymbol{u}, \boldsymbol{ui}$ *is a positive orthonormal basis of* P and therefore

$$\boldsymbol{v} = \boldsymbol{u}\cos\theta + \boldsymbol{ui}\sin\theta.$$

From this it follows, taking into account that $\boldsymbol{u}$ and $\boldsymbol{ui}$ are orthogonal, that

$$\boldsymbol{u} \wedge \boldsymbol{v} = \boldsymbol{u} \wedge \boldsymbol{ui}\sin\theta = \boldsymbol{uui}\sin\theta = \boldsymbol{i}\sin\theta,$$

and consequently

$$R = \boldsymbol{vu} = \boldsymbol{v} \cdot \boldsymbol{u} + \boldsymbol{v} \wedge \boldsymbol{u} = \cos\theta - \boldsymbol{i}\sin\theta = e^{-\boldsymbol{i}\theta}.$$

In this way we arrive at what we will call *Euler's spinorial formula*:

$$\underline{R}(\boldsymbol{x}) = e^{-\boldsymbol{i}\theta}\boldsymbol{x}e^{\boldsymbol{i}\theta}. \tag{4.6}$$

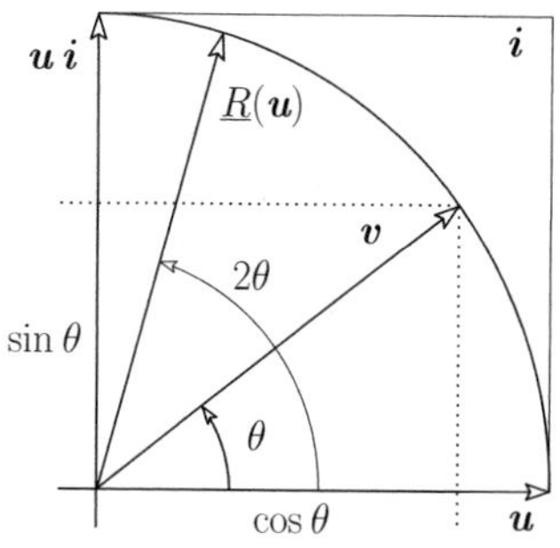

Fig. 4.1 Geometry of the rotor $R = \boldsymbol{vu}$, $\boldsymbol{u}$ and $\boldsymbol{v}$ linearly independent unit vectors

Now we have:

4.4.1 $\underline{R}$ *is a rotation in* P *(its axis is* $P^{\perp}$*) of amplitude* 2θ.

Proof A vector $\boldsymbol{x}$ orthogonal to P anticommutes with $\boldsymbol{u}$ and $\boldsymbol{v}$. Therefore, it commutes with $\boldsymbol{i}$, and $e^{-\boldsymbol{i}\theta}\boldsymbol{x}e^{\boldsymbol{i}\theta} = \boldsymbol{x}$. On the other hand, a vector $\boldsymbol{x} \in P$ anticommutes with $\boldsymbol{i}$ and hence $e^{-\boldsymbol{i}\theta}\boldsymbol{x}e^{\boldsymbol{i}\theta} = \boldsymbol{x}e^{2\boldsymbol{i}\theta}$, which is the rotation of $\boldsymbol{x}$ in P of amplitude $\alpha = 2\theta$ in the positive direction of P. □

4.4.2 (Remark) In the spinorial formula (4.6) what matters is the area element $\boldsymbol{a} = \boldsymbol{i}\alpha = 2\boldsymbol{i}\theta$, a bivector, not the separate factors $\boldsymbol{i}$ and α. Indeed, we can define $R_{\boldsymbol{a}} = e^{-\boldsymbol{a}/2}$ and then we have

$$\underline{R}_{\boldsymbol{a}}(\boldsymbol{x}) = e^{-\boldsymbol{a}/2}\boldsymbol{x}e^{\boldsymbol{a}/2}, \tag{4.6$'$}$$

which delivers the identity if $\boldsymbol{a} = 0$ and otherwise the rotation in the plane $P = |\boldsymbol{a}\rangle$ by the amplitude $\alpha = -\boldsymbol{a}\boldsymbol{i}$.

4.4.3 (Example: Rotations of E_3**)** Suppose $n = 3$. If $\boldsymbol{u}$ is a non-zero normal vector to the oriented plane P, its Hodge dual is $\boldsymbol{u}^* = \mathbf{i}\boldsymbol{u}$ ($\mathbf{i}$ the pseudoscalar of E_3) and

$$\underline{R}_{\boldsymbol{u}^*}(\boldsymbol{x}) = e^{-\mathbf{i}\boldsymbol{u}/2}\boldsymbol{x}e^{\mathbf{i}\boldsymbol{u}/2} \tag{4.7}$$

yields the rotation in the plane $|\boldsymbol{u}^*\rangle = \boldsymbol{u}^{\perp} = P$ by the angle $-\boldsymbol{u}^*\boldsymbol{i} = -|\boldsymbol{u}|\boldsymbol{i}^2 = |\boldsymbol{u}|$, where $\boldsymbol{i}$ is the oriented unit area of P. Indeed, it is enough to note that $\mathbf{i} = \boldsymbol{i}\boldsymbol{u}/|\boldsymbol{u}|$ and so $\boldsymbol{u}^*\boldsymbol{i} = \mathbf{i}\boldsymbol{u}\boldsymbol{i} = (\boldsymbol{i}\boldsymbol{u}/|\boldsymbol{u}|)\boldsymbol{u}\boldsymbol{i} = |\boldsymbol{u}|\boldsymbol{i}^2$. In other words, $\underline{R}_{\boldsymbol{u}^*}$ is the rotation $r_{\boldsymbol{u}}$ about $\boldsymbol{u}$ by an angle $|\boldsymbol{u}|$. Since Eq. (4.7) supplies the identity for $\boldsymbol{u} = \boldsymbol{0}$, it is natural to define $r_{\boldsymbol{0}} = \mathrm{Id}$.

Using (4.7) it is easy to deduce the *Olinde Rodrigues formulas* for the composition of two rotations (see, for example, [98, Ch. 1]).

4.4.4 (Effective Computation of Rotations) If $\boldsymbol{\omega}$ is a unit vector in E_3 and $\alpha \in \mathbb{R}$, the rotation $f = r_{\boldsymbol{\omega}\alpha}$ (which is also denoted $r_{\boldsymbol{\omega},\alpha}$), and its action on multivectors, can be efficiently computed using the formula (4.5) for the reflection of blades (of vectors in particular). In fact, $f = \underline{R}$, where the rotor R is the product $\boldsymbol{v}\boldsymbol{u}$ of any unit vectors $\boldsymbol{v}, \boldsymbol{u} \in \boldsymbol{\omega}^{\perp}$ such that $\boldsymbol{u} \cdot \boldsymbol{v} = \cos\frac{\alpha}{2}$ and the orientation of $\boldsymbol{u}, \boldsymbol{v}, \boldsymbol{\omega}$ has the same sign as α. Then $\underline{R} = \underline{\boldsymbol{v}}\,\underline{\boldsymbol{u}} = s_{\boldsymbol{v}}s_{\boldsymbol{u}} = m_{\boldsymbol{v}}m_{\boldsymbol{u}}$, which shows that applying $\underline{R}$ to any object is reduced to two successive reflections performed on the object.

If we are given $\boldsymbol{\omega}$ and α, E.4.7, p. 76, gives a practical recipe to find suitable unit vectors $\boldsymbol{u}$ and $\boldsymbol{v}$ such that the rotation associated with the rotor $\boldsymbol{v}\boldsymbol{u}$ is $r_{\boldsymbol{\omega},\alpha}$.

In the next chapter we will introduce the algebra of geometric quaternions (see 5.1.7, p. 83) and provide an account of how they relate to rotations. This question is also addressed, from a different perspective, in [98, 1.3.10 (Quaternions)].

4.5 Exercises

E.4.1 As in the last step of the proof of 4.1.7, let D a totally isotropic subspace of E and $f \in \mathrm{O}_q(E)$ be an isometry that leaves D invariant. If the restriction of f to D is the identity, and it induces the identity in E/D, then $f \in \mathrm{SO}_q(E)$.

Hint Extend a basis $\boldsymbol{e}_1, \ldots, \boldsymbol{e}_k$ of D to a basis $\mathbf{e} = \boldsymbol{e}_1, \ldots, \boldsymbol{e}_k, \boldsymbol{e}_{k+1}, \ldots, \boldsymbol{e}_n$ of E and let $\bar{\boldsymbol{e}}_j = \pi(\boldsymbol{e}_j)$ for $j = k+1, \ldots, n$, where $\pi : E \to E/D$ is the quotient map. Show that there exists a $k \times (n-k)$ matrix A such that the matrix of f with respect to $\mathbf{e}$ has the form $\begin{pmatrix} I_k & A \\ 0 & I_{n-k} \end{pmatrix}$.

E.4.2 (Lipschitz Versors and the Lipschitz Representation) A *Lipschitz versor* is an invertible multivector v such that $vEv^{-1} = E$ (it is like the definition of versor, but not requiring that it is even or odd). The purpose of the exercise is to study the set $\Gamma_{r,s}$ of Lipschitz versors and its relation to the versor group $\mathcal{V}_{r,s}$. Let us break the task in several subtasks, with hints for the less obvious.

(1) $\Gamma_{r,s}$ is a subgroup of $\mathcal{G}^\times$.
(2) $\mathcal{V}_{r,s}$ is a subgroup of $\Gamma_{r,s}$.
(3) For any $v \in \Gamma_{r,s}$, the map $\rho_v : E \to E$ defined by $\rho_v(\boldsymbol{x}) = v\boldsymbol{x}v^{-1}$ is an isometry.
(4) The map $\rho : \Gamma_{r,s} \to \mathrm{O}_{r,s}$, $v \mapsto \rho_v$, is a group homomorphism.
(5) $\ker(\rho) = \mathbb{R}^\times$.
(6) $\Gamma_{r,s} = \mathcal{V}_{r,s}$.

Hints In (1) and (4), use that $(vw)^{-1} = w^{-1}v^{-1}$ for any $v, w \in \mathcal{G}^\times$. In (2), use that $\hat{v} = \pm v$ for any $v \in \mathcal{V}_{r,s}$. To prove (5), reuse the proof that $\ker(\tilde{\rho}) = \mathbb{R}^\times$.

A proof of (6) can be structured as follows. Given $w \in \Gamma_{r,s}$, pick $v \in \mathcal{V}_{r,s}$ such that $\tilde{\rho}(v) = \rho(w)$ (it exists because the image of $\tilde{\rho}$ is $\mathrm{O}_{r,s}$). Since $\tilde{\rho}(v) = \pm\rho(v)$, we have $\rho(wv^{-1}) = \pm 1$. If $\rho(wv^{-1}) = 1$, then $\lambda := wv^{-1} \in \mathbb{R}^\times$ (by (5)) and $w = \lambda v \in \mathcal{V}_{r,s}$. If $\rho(u) = -1$, where $u = wv^{-1}$, then we have $u\boldsymbol{x} = -\boldsymbol{x}u$ for all vectors $\boldsymbol{x}$, which by 3.3.7 can happen only if u is a pseudoscalar. Thus $w = uv$, which belongs to $\mathcal{V}_{r,s}$.

Remark The reason why ρ cannot play the role of $\tilde{\rho}$ is that for any $\boldsymbol{e}, \boldsymbol{x} \in E$, $\boldsymbol{e}$ non-null, $\rho_{\boldsymbol{e}}(\boldsymbol{x}) = \boldsymbol{e}\boldsymbol{x}\boldsymbol{e}^{-1} = s_{\boldsymbol{e}}(\boldsymbol{x})$, while $\tilde{\rho}_{\boldsymbol{e}}(\boldsymbol{x}) = \hat{\boldsymbol{e}}\boldsymbol{x}\boldsymbol{e}^{-1} = m_{\boldsymbol{e}}(\boldsymbol{x})$. Note in particular that if n is odd, then the image of ρ is contained in $\mathrm{SO}_{r,s}$.

E.4.3 In the example 4.2.3 we have shown that $\underline{I} = -\mathrm{Id}$. Prove the same identity using an orthonormal basis $\boldsymbol{e}_1, \ldots, \boldsymbol{e}_n$ such that $I = \boldsymbol{e}_1 \cdots \boldsymbol{e}_n$.

Hint Use that $\underline{I} = m_{\boldsymbol{e}_1} \cdots m_{\boldsymbol{e}_n}$ to see that $\underline{I}(\boldsymbol{e}_j) = -\boldsymbol{e}_j$ for any $j = 1, \ldots, n$.

E.4.4 For the Euclidean plane E_2, show that any axial symmetry is a reflection and vice versa. Is the same statement true for the Euclidean space E_n, $n \geqslant 3$?

E.4.5 Show that in the Euclidean plane E_2 and the Euclidean space E_3 any rotation is the composition of two reflections. Show that in E_3 an improper isometry is either $-\mathrm{Id}$, which is the composition of three reflections, or a reflection.

E.4.6 If v is a versor of E_3 and R is a rotor such that $\underline{R} = r_{\boldsymbol{\omega},\alpha}$ (the rotation about the unit vector $\boldsymbol{\omega}$ by an angle α), then $S = \underline{v}(R)$ is a rotor. Show that the rotation $\underline{S}$ is equal to $r_{\boldsymbol{\omega}',\alpha}$, where $\boldsymbol{\omega}' = \underline{v}(\boldsymbol{\omega})$.

E.4.7 Let $\boldsymbol{\omega}$ be a unit vector of the Euclidean space E_3 and $\alpha \in \mathbb{R}$. Let $\boldsymbol{e}_1, \boldsymbol{e}_2, \boldsymbol{e}_3$ be a positively oriented orthonormal basis of E_3. Define vectors $\boldsymbol{u}_1, \boldsymbol{u}_2$ as follows. If $\boldsymbol{\omega} = \boldsymbol{e}_3$, take $\boldsymbol{u}_1 = \boldsymbol{e}_1$ and $\boldsymbol{u}_2 = \boldsymbol{e}_2$. Otherwise, if $\boldsymbol{\omega} = w_1\boldsymbol{e}_1 + w_2\boldsymbol{e}_2 + w_3\boldsymbol{e}_3$, take

$$\boldsymbol{u}_1 = (-w_2\boldsymbol{e}_1 + w_1\boldsymbol{e}_2)/\rho \ \text{ and } \ \boldsymbol{u}_2 = (-w_1w_3\boldsymbol{e}_1 - w_2w_3\boldsymbol{e}_2 + \rho^2\boldsymbol{e}_3)/\rho,$$

where $\rho = +(w_1^2 + w_2^2)^{1/2}$.

(1) Show that $\boldsymbol{u}_1, \boldsymbol{u}_2, \boldsymbol{\omega}$ is a positively oriented orthonormal basis of E_3.
(2) If we set $\boldsymbol{u} = \boldsymbol{u}_1$ and $\boldsymbol{v} = \boldsymbol{u}_1 \cos\frac{\alpha}{2} + \boldsymbol{u}_2 \sin\frac{\alpha}{2}$, the rotation $r_{\boldsymbol{\omega},\alpha}$ is the composition of reflections $m_{\boldsymbol{v}}m_{\boldsymbol{u}}$.
(3) If $w_3 = 0$, check that $\boldsymbol{u}_2 = \boldsymbol{e}_3$.

Chapter 5
Zooming in on Rotor Groups

This chapter is devoted to a closer study of the rotor group $\mathcal{R} = \mathcal{R}_{r,s}$, and as a byproduct of its primacy (in the sense given to this expression in the last chapter), also of the other spinorial and orthogonal groups.

We begin with a few detailed examples in low dimensions (Sect. 5.1), a task that is facilitated by a simple characterization of $\mathcal{R}$, namely 5.1.1, in dimension not greater than 5. To note the examples in dimensions 1 and 2, as they have specific properties that do not appear in higher dimensions, and also in dimension 3, including the always surprising case of the Euclidean space E_3 and the isomorphisms $\mathcal{R}_{2,1} \simeq \mathcal{R}_{1,2} \simeq \mathrm{SL}_2$. In Sect. 5.2, we define and study plane rotors in any dimension and signature, and prove that they generate $\mathcal{R}$. In fact, we show that any rotor is the product of at most $n/\!\!/2$ plane rotors. This result is fundamental for applications, as for example in projective geometry (▷ 6.4.2) and several other applications in geometry and physics (▷ 6.4.4).

The focus of Sects. 5.3 and 5.4 are the notions of infinitesimal rotation and infinitesimal rotor, respectively. The infinitesimal rotations turn out to be the skew-symmetric endomorphisms of (E, q), that is, endomorphisms h such that

$$(h\boldsymbol{x}) \cdot \boldsymbol{y} + \boldsymbol{x} \cdot (h\boldsymbol{y}) = 0 \text{ for all } \boldsymbol{x}, \boldsymbol{y} \in E, \tag{5.1}$$

or, equivalently, such that $(h\boldsymbol{x}) \cdot \boldsymbol{x} = 0$ for all $\boldsymbol{x} \in E$ (see E.5.3, p. 103). Similarly, the infinitesimal rotors turn out to be the space $\mathcal{G}^2$ of bivectors. Now, by general principles (▷ 6.5.9), $\mathcal{G}^2$ and the space $\mathfrak{so}(E, q)$ of skew-symmetric endomorphisms of (E, q) should be isomorphic. This fact is established in a strong sense that will be spelled out in a precise form. This isomorphism amounts to the infinitesimal version of the 2:1 group homomorphism $\rho : \mathcal{R} \to \mathrm{SO}^0$. We end Sect. 5.4 with a short account of the structure of normal endomorphisms of the Euclidean space E_n (which include symmetric, skew-symmetric, and orthogonal endomorphisms) and use the result to show that any isometry of E_n is the exponential of a skew-symmetric endomorphism.

S. Xambó-Descamps, *Real Spinorial Groups*, SpringerBriefs in Mathematics,
https://doi.org/10.1007/978-3-030-00404-0_5

The techniques used in this chapter include, in addition to the algebraic tools used so far, a few notions of a topological and differentiable character. In our context, these ideas should be widely accessible, for all the groups are subsets of real vector spaces and therefore we can readily use the tools of multivariate calculus. Nevertheless, we refresh the most basic in Sect. 6.1, the more advanced (concerning differentiable maps, manifolds, and Lie groups) in Sect. 6.5, and all along we take care to insert more specific pointers and remarks where they may be more useful.

5.1 Examples of Rotor Groups

Let us begin with a result that facilitates the description of the rotor groups for $n \leqslant 5$, as it is a simple condition imposed on even multivectors.

5.1.1 *If* $n \leqslant 5$, $\mathcal{R} = \{R \in \mathcal{G}^+ \mid R\tilde{R} = 1\}$.

Proof Cf. [106, Prop. 6.20]. It is obvious that $\{R \in \mathcal{G}^+ \mid R\tilde{R} = 1\}$ contains $\mathcal{R}$. To see the converse inclusion, let R be an even multivector such that $R\tilde{R} = 1$. To prove that $R \in \mathcal{R}$, it is sufficient to see, by 4.2.2, that $y = R\boldsymbol{x}\tilde{R}$ is a vector for any vector $\boldsymbol{x}$.

Indeed, since y is odd and $\tilde{y} = y$, we have $y = y_1 + y_5$ (as $n \leqslant 5$ and $y_3 = 0$ because $\tilde{y}_3 = -y_3$). Thus, it will be enough to check that $y_5 = 0$. Since this is trivial for $n < 5$, we may assume that $n = 5$. In this case, $y = y_1 + \lambda I$, where I is a pseudoscalar. Using that I is a central element (as $n = 5$ is odd) in the third step below, and the alternative expression for the metric on $\mathcal{G}$ (see 3.3.9) for the commutation in the fourth step, we have:

$$\lambda = \left(yI^{-1}\right)_0 = \left(R\boldsymbol{x}\tilde{R}I^{-1}\right)_0 = \left(R\boldsymbol{x}I^{-1}\tilde{R}\right)_0 = \left(\boldsymbol{x}I^{-1}\tilde{R}R\right)_0 = \left(\boldsymbol{x}I^{-1}\right)_0 = 0,$$

and this completes the proof. □

5.1.2 (Remark) In E.5.4, p. 103, you will discover an even element R in $\mathcal{G}_6$ such that $R\tilde{R} = 1$ and such that $R\boldsymbol{e}_1\tilde{R}$ is a 5-blade. Thus, 5.1.1 does not hold for $n > 5$.

5.1.3 (Topological Remarks) The notions of open and closed sets, of continuous and differentiable functions and maps, are used as in multi-variable calculus relative to any real vector space V. For example, a (continuous) *path* on a subset X of V is a continuous map $\gamma : [0, 1] \to X$. The points $\gamma(0)$ and $\gamma(1)$ are the *end-points* of the path, and we say that these points are (path) connected, or that $\gamma(0)$ can be (path) connected to $\gamma(1)$. If any two points of X can be connected, we say that X is *path-connected*. Any path-connected set X is *connected*, meaning that it is not the disjoint union of two non-empty closed (or open) sets of X. This is an easy consequence of the fact that the interval $[0, 1] \subset \mathbb{R}$ is connected. The converse is only true if in addition X is *locally path-connected*, meaning that any point of X can be connected to all points of one of its open neighborhoods. See E.5.5, p. 103, for further details about these facts. Since we are going to deal only with locally path-connected sets, in this chapter we will use the terms connected and path-connected as synonymous.

A connected set X is said to be *simply connected* if it is connected and any closed path:

$$\gamma : [0, 1] \to X, \quad \gamma(0) = \gamma(1),$$

can be continuously shrunk to a point without leaving X. For example, the n-dimensional sphere S^n is simply connected for all $n \geqslant 2$, while S^1 (a circle) is connected but not simply connected (the closed path consisting in going around S^1 once cannot be shrunk to a point without leaving S^1). The zero-dimensional sphere consists of two points, and hence it is not connected (▷ 6.1.1 and ▷ 6.1.4 for more precise definitions and arguments).

In the examples that follow, we set $\mathcal{P} = \mathcal{P}_{r,s} = \mathrm{Pin}_{r,s}$, $\mathcal{S} = \mathcal{S}_{r,s} = \mathrm{Spin}_{r,s}$, so that $\mathcal{R}_{r,s} = \mathcal{S}^+_{r,s}$ (see 4.3.4). In particular, $\mathcal{P}_n = \mathcal{P}_{n,0}$, $\mathcal{S}_n = \mathcal{S}_{n,0}$, and $\mathcal{R}_n = \mathcal{R}_{n,0}$. For the anti-Euclidean signature, we will write $\mathcal{P}_{\bar{n}} = \mathcal{P}_{0,n}$, and similarly $\mathcal{S}_{\bar{n}} = \mathcal{S}_{0,n}$, $\mathcal{R}_{\bar{n}} = \mathcal{S}_{0,n}$. The same conventions are made to denote, when required, the signature for the space E and the groups O and SO.

5.1.4 (Rotors of E_1 and $E_{\bar{1}}$) For $n = 1$, the possible signatures are $1 \sim (1, 0)$ and $\bar{1} \sim (0, 1)$, and we have (see Fig. 5.1):

$$\mathcal{S}_1 = \mathcal{R}_1 = \mathcal{S}_{\bar{1}} = \mathcal{R}_{\bar{1}} = \{\pm 1\} \simeq \mathbb{Z}_2,$$

because by 5.1.1 the group $\mathcal{R}_1$ consists of the scalars λ such that $\lambda^2 = 1$. Note that it is not connected. On the other hand,

$$\mathcal{P}_1 = \mathcal{P}_{\bar{1}} = \{\pm 1, \pm \boldsymbol{u}\},$$

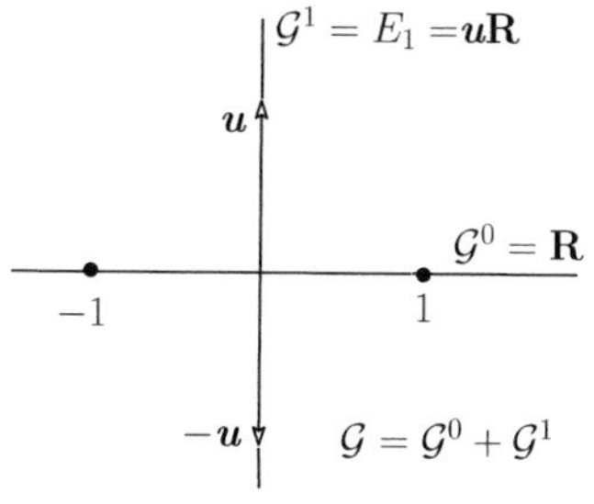

Fig. 5.1 The rotor group of E_1 and $E_{\bar{1}}$ is $\{\pm 1\}$, which is disconnected. The pinor group $\mathcal{P}$ is $\{\pm 1, \pm \boldsymbol{u}\}$, where $\boldsymbol{u} \in E_1$ is a unit vector. In the Euclidean case, $\boldsymbol{u}^2 = 1$ and we have a group isomorphic to $\mathbb{Z}_2 \times \mathbb{Z}_2$. In the anti-Euclidean case, $\boldsymbol{u}^2 = -1$ and we have a cyclic group of order 4 generated by $\boldsymbol{u}$

where $\boldsymbol{u} \in E$ is any given unit vector. If $\boldsymbol{u}^2 = -1$ (which happens in the case $E_{\bar{1}}$), we have the group:

$$\{1, \boldsymbol{u}, -1, -\boldsymbol{u}\} = \{1, \boldsymbol{u}, \boldsymbol{u}^2, \boldsymbol{u}^3\} \simeq \mathbb{Z}_4,$$

and if $\boldsymbol{u}^2 = 1$ (which happens in the case E_1), then the group is isomorphic to:

$$\mathbb{Z}_2 \times \mathbb{Z}_2 = \{(0, 1), (1, 0), (0, 1), (1, 1)\},$$

as $\boldsymbol{u}^2 = (-1)^2 = (-\boldsymbol{u})^2 = 1$. These conclusions may be summarized as $\mathcal{P}_1 \simeq \mathbb{Z}_2 \times \mathbb{Z}_2$ (the *Klein group* of order 4) in the Euclidean case and $\mathcal{P}_1 \simeq \mathbb{Z}_4$ (the cyclic group of order 4) in the anti-Euclidean case.

Finally, $\mathrm{O}_1 = \{\pm\mathrm{Id}\}$ and $\mathrm{SO}_1 = \{\mathrm{Id}\}$, with $\tilde{\rho}(\pm 1) = \mathrm{Id}$ and $\tilde{\rho}(\pm\boldsymbol{u}) = -\mathrm{Id}$.

5.1.5 (Rotors of E_2) The spaces E_2 and $E_{\bar{2}}$ have a similar treatment. If $\boldsymbol{i}$ is the area unit, $\mathcal{G}^+ = \{\alpha + \beta\boldsymbol{i} \mid \alpha, \beta \in \mathbb{R}\} = \mathbf{C}$ (the geometric complex numbers or complex scalars, cf. [98]). Since $\boldsymbol{i}^2 = -1$ and $\tilde{\boldsymbol{i}} = -\boldsymbol{i}$, $(\alpha + \beta\boldsymbol{i})(\alpha + \beta\boldsymbol{i})^{\sim} = \alpha^2 + \beta^2$ and

$$\mathcal{S}_2 = \mathcal{R}_2 = \mathcal{S}_{\bar{2}} = \mathcal{R}_{\bar{2}} = \left\{\alpha + \beta\boldsymbol{i} \in \mathcal{G}^+ \mid \alpha^2 + \beta^2 = 1\right\} = U_1,$$

where $U_1 = \{e^{\boldsymbol{i}\theta} : 0 \leqslant \theta < 2\pi\}$ denotes the unit circle in $\mathbf{C}$ (see Fig. 5.2a). This group is connected, but not simply connected (a turn of the unit circle cannot be shrinked to 1, or, more precisely, the *fundamental group* of U_1 is $\mathbb{Z}$, $\pi_1(U_1) \simeq \mathbb{Z}$ in symbols (▷ 6.1.4 for more details).

The isometry $\tilde{\rho}(e^{-\boldsymbol{i}\theta})$ is the counterclockwise rotation (in relation to the orientation $\boldsymbol{i}$) by an angle 2θ. Indeed, in this case $\boldsymbol{i}$ anticommutes with vectors and hence $\tilde{\rho}(e^{-\boldsymbol{i}\theta})(\boldsymbol{x}) = e^{-\boldsymbol{i}\theta}\boldsymbol{x}e^{\boldsymbol{i}\theta} = \boldsymbol{x}e^{2\boldsymbol{i}\theta}$, which is the result of rotating $\boldsymbol{x}$ by an angle 2θ. In this case, SO_2 is also isomorphic to U_1, but the map $\tilde{\rho} : U_1 \to U_1 \simeq \mathrm{SO}_2$

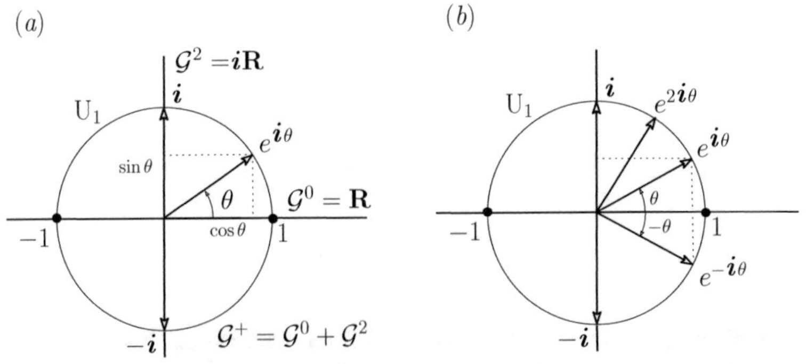

Fig. 5.2 The rotor group of E_2 and $E_{\bar{2}}$ is the circle $\mathcal{R}_2 = \{e^{\boldsymbol{i}\theta} \mid 0 \leqslant \theta < 2\pi\} = U_1$ in $\mathcal{G}^+ = \mathbf{C}$. The group SO_2 is also isomorphic to U_1, but the 2:1 group homomorphism $\mathcal{R}_2 \to \mathrm{SO}_2$ becomes the map $e^{-\boldsymbol{i}\theta} \mapsto e^{2\boldsymbol{i}\theta}$, so that going around the rotor U_1 once in a clockwise sense produces going around the rotation U_1 twice in a counterclockwise sense

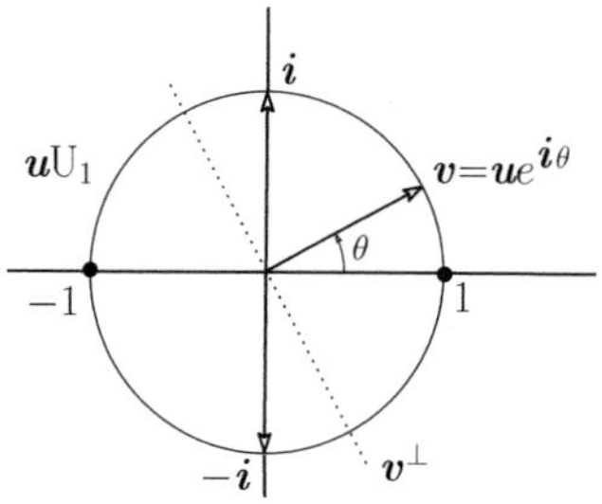

Fig. 5.3 Fix the vector $\boldsymbol{u}$, say $\boldsymbol{u} = \boldsymbol{e}_1$. The circle $\boldsymbol{u}U_1 \subset E_2$ provides a 2:1 representation of the set O_2^- of reflections of E_2, as $\tilde{\rho}(\boldsymbol{v})$ is the reflection along $\boldsymbol{v}$, or across the line $\boldsymbol{v}^{\perp}$, and $\pm\boldsymbol{v}$ yield the same reflection

is given by $e^{-i\theta} \mapsto e^{2i\theta}$. Note that when θ varies from 0 to 2π, $e^{-i\theta}$ goes around U_1 once, in a clockwise sense, while $e^{2i\theta}$ goes around U_1 twice, thus providing, in this case, a nice visualization of the meaning of the $2:1$ onto homomorphism $\mathcal{R}_2 \to \mathrm{SO}_2$ (see Fig. 5.2b).

If $\boldsymbol{u}$ is any given unit vector, say $\boldsymbol{u} = \boldsymbol{e}_1$, then

$$\mathcal{P}_2 = \mathcal{P}_{\bar{2}} = \mathcal{R}_2 \sqcup \boldsymbol{u}\mathcal{R}_2 = U_1 \sqcup \boldsymbol{u}U_1,$$

and $\tilde{\rho}(\boldsymbol{u}e^{\boldsymbol{i}\theta})$ is the reflection along the direction $\boldsymbol{v} = \boldsymbol{u}e^{\boldsymbol{i}\theta}$, for $-\boldsymbol{v}\boldsymbol{v}\boldsymbol{v}^{-1} = -\boldsymbol{v}$. Thus, we see that topologically $\mathcal{P}_2 = \mathcal{P}_{\bar{2}}$ is the disjoint union of two circles: $U_1 \subset \mathbf{C} = \mathcal{G}^+$ and $\boldsymbol{u}U_1 \subset \mathcal{G}^- = E_2$. The 2:1 map $\boldsymbol{u}U_1 \to \mathrm{O}_2^-$ is manifested in the fact that $\boldsymbol{v}$ and $-\boldsymbol{v}$ give the same reflection. In other words, topologically O_2^- is also a circle that is circled twice when going once around the circle $\boldsymbol{u}U_1$ (see Fig. 5.3).

5.1.6 (Rotors of $E_{1,1}$) Let $\boldsymbol{e}_0, \boldsymbol{e}_1$ be an orthonormal basis of $E_{1,1}$ and $\boldsymbol{i} = \boldsymbol{e}_1\boldsymbol{e}_0$ (in Lorentzian spaces, i.e., spaces of signature $(1, n-1)$, it is customary to set the "temporal axis" $\boldsymbol{e}_0$ as the "vertical direction"). We still have $\mathcal{G}^+ = \{\alpha + \beta\boldsymbol{i} \mid \alpha, \beta \in \mathbb{R}\}$ and $(\alpha + \beta\boldsymbol{i})^{\sim} = \alpha - \beta\boldsymbol{i}$, but

$$(\alpha + \beta\boldsymbol{i})(\alpha + \beta\boldsymbol{i})^{\sim} = \alpha^2 - \beta^2$$

inasmuch as $\boldsymbol{i}^2 = 1$. Thus, the rotor group is

$$\mathcal{R}_{1,1} = \{\alpha + \beta\boldsymbol{i} \mid \alpha^2 - \beta^2 = 1\},$$

which shows that it has two connected (and simply connected) components: the two branches of a hyperbola in $\mathcal{G}^+$. These branches are distinguished by the sign of α (Fig. 5.4a) and are parameterized by $\alpha = \epsilon \cosh\lambda$, $\beta = \sinh\lambda$ ($\epsilon = \pm 1$, $\lambda \in \mathbb{R}$).

After a few calculations (in which we use that $\boldsymbol{i}$ anticommutes with $\boldsymbol{e}_0$ and $\boldsymbol{e}_1$, and basic properties of cosh and sinh), we find that the action of

$$R = R_{\epsilon,\lambda} = \epsilon\cosh\lambda + \boldsymbol{i}\sinh\lambda = \epsilon e^{\epsilon\lambda\boldsymbol{i}}$$

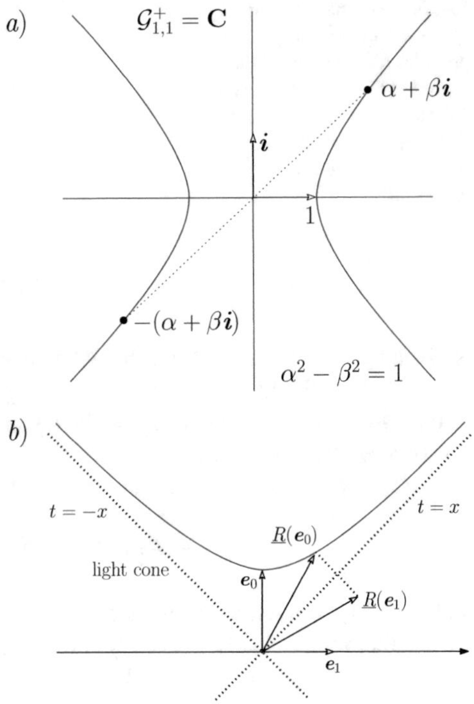

Fig. 5.4 Geometric aspects of the $E_{1,1}$ rotors. *a*) The two components of the rotor group $\mathcal{R}_{1,1}$. *b*) Action of a rotor R on vectors. The "light cone" is the physically motivated name for the isotropic cone $q(\boldsymbol{x}) = 0$, which in terms of the components of $\boldsymbol{x}$ with respect to $\boldsymbol{e}_1$ and $\boldsymbol{e}_0$, say $\boldsymbol{x} = x\boldsymbol{e}_1 + t\boldsymbol{e}_0$, is given by the equation $t^2 - x^2 = 0$ and so it is composed of the lines $t = \pm x$

on a vector $\boldsymbol{x}$ is given by:

$$\underline{R}(\boldsymbol{x}) = R\boldsymbol{x}\tilde{R} = \boldsymbol{x}e^{-\epsilon 2\lambda \boldsymbol{i}}.$$

In particular, we get, using that $\boldsymbol{e}_0\boldsymbol{i} = -\boldsymbol{e}_1$ and $\boldsymbol{e}_1\boldsymbol{i} = -\boldsymbol{e}_0$, the following expressions (see Fig. 5.4b):

$$\underline{R}(\boldsymbol{e}_0) = \boldsymbol{e}_0 \cosh 2\epsilon\lambda + \boldsymbol{e}_1 \sinh 2\epsilon\lambda, \quad \underline{R}(\boldsymbol{e}_1) = \boldsymbol{e}_0 \sinh 2\epsilon\lambda + \boldsymbol{e}_1 \cosh 2\epsilon\lambda.$$

We see that $R_{\epsilon,\lambda}$ and $R_{-\epsilon,-\lambda}$ produce the same isometry (as they must, because $R_{-\epsilon,-\lambda} = -R_{\epsilon,\lambda}$ and $\pm R$ yield the same isometry), and consequently $\mathrm{SO}_{1,1}^0$ is isomorphic to (the additive group) $\mathbb{R}$ via the map (in matrix form):

$$t \mapsto H_t = \begin{pmatrix} \cosh 2t & \sinh 2t \\ \sinh 2t & \cosh 2t \end{pmatrix}.$$

Thus, we have seen that $\mathcal{R}_{1,1}$ has two connected components, which implies that $\mathcal{P}_{1,1}$ has eight connected components. Indeed, if we set $\mathcal{R}^+ = \{R_{1,\lambda}\}$ and $\mathcal{R}^- = \{R_{-1,\lambda}\}$, these components are

$$\mathcal{R}^+,\ \mathcal{R}^-,\ \boldsymbol{e}_0\mathcal{R}^+,\ \boldsymbol{e}_0\mathcal{R}^-,\ \boldsymbol{e}_1\mathcal{R}^+,\ \boldsymbol{e}_1\mathcal{R}^-,\ \boldsymbol{i}\mathcal{R}^+,\ \boldsymbol{i}\mathcal{R}^-.$$

From this, it follows that $\mathrm{O}_{1,1}$ has four connected components:

$$\mathrm{SO}^0_{1,1},\ m_{e_0}\mathrm{SO}^0_{1,1},\ m_{e_1}\mathrm{SO}^0_{1,1},\ \text{and}\ m_{e_1}m_{e_0}\mathrm{SO}^0_{1,1} = -\mathrm{SO}^0_{1,1}.$$

5.1.7 (Rotors of E_3) Let $\boldsymbol{e}_1, \boldsymbol{e}_2, \boldsymbol{e}_3$ be an orthonormal basis of E_3, $\boldsymbol{i} = \boldsymbol{e}_1\boldsymbol{e}_2\boldsymbol{e}_3$ (unit volume, which satisfies $\boldsymbol{i}^2 = -1$ and $\tilde{\boldsymbol{i}} = -\boldsymbol{i}$), and

$$\mathbf{H} = \mathcal{G}^+ = \{\sigma + \boldsymbol{w}\boldsymbol{i} \mid \sigma \in \mathbb{R}, \boldsymbol{w} \in E_3\}$$

the even geometric algebra. We know that $\mathbf{H}$ is a Euclidean space (with the multivector metric induced by the metric of E_3, both denoted by q). In fact, this is confirmed in this case by the expression:

$$q(\sigma + \boldsymbol{w}\boldsymbol{i}) = (\sigma + \boldsymbol{w}\boldsymbol{i})(\sigma + \boldsymbol{w}\boldsymbol{i})^{\sim} = \sigma^2 + |\boldsymbol{w}|^2,$$

which manifestly shows that q is positive definite. Thus, $\mathbf{H}$ is a (skew) field, as the inverse of $h = \sigma + \boldsymbol{w}\boldsymbol{i} \neq 0$ is $h^{-1} = \tilde{h}/q(h)$, and consequently

$$\mathcal{S}_3 = \mathcal{S}_{\tilde{3}} = \mathcal{R}_3 = \mathcal{R}_{\tilde{3}} = \left\{\sigma + \boldsymbol{w}\boldsymbol{i} \in \mathbf{H} \mid \sigma^2 + |\boldsymbol{w}|^2 = 1\right\} = \{h \in \mathbf{H} \mid q(h) = 1\}.$$

This shows that $\mathcal{S}_3 = \mathcal{R}_3$ is the unit sphere of $\mathbf{H}$ and hence that it is connected and simply connected (▷ 6.1.4). Then, the double cover

$$\rho : \mathcal{R}_3 \to \mathrm{SO}_3$$

yields that SO_3 is connected. Note that for $h \in \mathcal{R}_3$ we have $\tilde{h} = h^{-1}$ and hence h acts on E_3 as the rotation

$$\rho_h : \boldsymbol{x} \mapsto h\boldsymbol{x}h^{-1}.$$

5.1.8 (Exponential Form of $h \in \mathcal{R}_3$)

(1) *Given $h = \sigma + \boldsymbol{w}\boldsymbol{i} \in \mathcal{R}_3$, $h \neq \pm 1$, there is a unique $\varphi \in (0, \pi)$ such that $\sigma = \cos\varphi$ and $|\boldsymbol{w}| = \sin\varphi$.*
(2) *If we set $\boldsymbol{\omega} = \boldsymbol{w}/|\boldsymbol{w}| = \boldsymbol{w}/\sin\varphi$, then $h = \cos\varphi + \boldsymbol{\omega}\boldsymbol{i}\sin\varphi = e^{\boldsymbol{\omega}\boldsymbol{i}\varphi}$.*
(3) *We have $\rho_h = r_{\boldsymbol{\omega},-2\varphi}$* (the rotation about $\boldsymbol{\omega}$ through the angle -2φ).
(4) *Let $\boldsymbol{u}$ and $\boldsymbol{v}$ be linearly independent unit vectors, and let $\alpha \in (0, \pi)$ be defined by $\boldsymbol{u} \cdot \boldsymbol{v} = \cos\alpha$. Then, the exponential form of the rotor $h = \boldsymbol{v}\boldsymbol{u}$ is $e^{-\boldsymbol{\omega}\boldsymbol{i}\alpha}$, where $\boldsymbol{\omega} = (\boldsymbol{u} \times \boldsymbol{v})/|\boldsymbol{u} \times \boldsymbol{v}|$. In particular, $\rho_h = r_{\boldsymbol{\omega},2\alpha}$.*

Proof

(1) This is clear from the equality $1 = h\tilde{h} = \sigma^2 + \boldsymbol{w}^2 = \sigma^2 + |\boldsymbol{w}|^2$ and the fact that $|\boldsymbol{w}| > 0$ (for otherwise $h = \sigma = \pm 1$).
(2) The vector $\boldsymbol{\omega} = \boldsymbol{w}/\sin(\varphi) = \boldsymbol{w}/|\boldsymbol{w}|$ is a unit vector and we have $\boldsymbol{w} = \boldsymbol{\omega}\sin\varphi$. Therefore, $h = \cos\varphi + \boldsymbol{\omega}\boldsymbol{i}\sin\varphi = e^{\boldsymbol{\omega}\boldsymbol{i}\varphi}$ because $(\boldsymbol{\omega}\boldsymbol{i})^2 = -1$.

(3) Indeed, $\rho_h(\boldsymbol{x}) = e^{\boldsymbol{\omega i}\varphi}\boldsymbol{x}e^{-\boldsymbol{\omega i}\varphi}$. Since $\boldsymbol{\omega}$ commutes with h, it is clear that $\rho_h(\boldsymbol{\omega}) = \boldsymbol{\omega}$. On the other side, if $\boldsymbol{x}$ is orthogonal to $\boldsymbol{\omega}$, then $\boldsymbol{x}$ anticommutes with h and therefore $\rho_h(\boldsymbol{x}) = \boldsymbol{x}e^{-\boldsymbol{\omega i}2\varphi}$ which is the result of rotating $\boldsymbol{x}$ by an angle -2φ in the plane $|\boldsymbol{\omega i}\rangle = \boldsymbol{\omega}^{\perp}$.

(4) We have $h = \boldsymbol{vu} = \boldsymbol{v}\cdot\boldsymbol{u} - \boldsymbol{u}\wedge\boldsymbol{v} = \cos\alpha - \boldsymbol{wi}$, where $\boldsymbol{w} = -(\boldsymbol{u}\wedge\boldsymbol{v})\boldsymbol{i} = \boldsymbol{u}\times\boldsymbol{v}$. Since $q(h) = 1$, $q(\boldsymbol{w}) = q(\boldsymbol{wi}) = 1 - \cos^2\alpha = \sin^2\alpha$, so $\boldsymbol{\omega} = \boldsymbol{w}/\sin\alpha$ is a unit vector and $h = \cos\alpha - \boldsymbol{\omega i}\sin\alpha = e^{-\boldsymbol{\omega i}\alpha}$. □

For example, if we let h be one of the three rotors

$$\boldsymbol{i}_1 = \boldsymbol{e}_2\boldsymbol{e}_3,\ \boldsymbol{i}_2 = \boldsymbol{e}_1\boldsymbol{e}_3,\ \boldsymbol{i}_3 = \boldsymbol{e}_1\boldsymbol{e}_2,$$

the isometries we get are the axial symmetries whose axes are $\boldsymbol{e}_1, \boldsymbol{e}_2, \boldsymbol{e}_3$, respectively. It is immediate to check that $\boldsymbol{i}_1^2 = \boldsymbol{i}_2^2 = \boldsymbol{i}_3^2 = \boldsymbol{i}_1\boldsymbol{i}_2\boldsymbol{i}_3 = -1$, which show that the field **H** is isomorphic to the field $\mathbb{H}$ of Hamilton's quaternions (mapping $\boldsymbol{i}_1, \boldsymbol{i}_2, \boldsymbol{i}_3$ to $\boldsymbol{i}, \boldsymbol{j}, \boldsymbol{k}$). Owing to the geometric substance of **H**, we will say that **H** is the field of *geometric quaternions*.

5.1.9 (Remark) The two components of $\mathcal{P}_3$ are $\mathcal{R}_3$ and $\boldsymbol{u}\mathcal{R}_3$, where $\boldsymbol{u}$ is any given unit vector, and the two components of O_3 are SO_3 and $m_{\boldsymbol{u}}\mathrm{SO}_3$.

If f is a rotation, $m_{\boldsymbol{u}}f$ is $-\mathrm{Id}$ in case $f = -m_{\boldsymbol{u}} = s_{\boldsymbol{u}}$, and otherwise it is a reflection, say $m_{\boldsymbol{v}}$. To determine $\boldsymbol{v}$, let $f = r_{\boldsymbol{\omega},\alpha}$ and $R = e^{-\boldsymbol{\omega i}\alpha/2}$, so that $f = \underline{R}$. Then:

$$m_{\boldsymbol{u}}f = \tilde{\rho}(\boldsymbol{u})\tilde{\rho}(R) = \tilde{\rho}(\boldsymbol{u}R).$$

If $\boldsymbol{u}R$ were a vector, we would have $m_{\boldsymbol{u}}f = m_{\boldsymbol{u}R}$, the reflection in the direction of $\boldsymbol{u}R$. But in general, $\boldsymbol{u}R$ is the sum of a vector and a pseudoscalar, so we have to work a bit more. If we apply $m_{\boldsymbol{u}}f$ to $\boldsymbol{u}R$, we get $-\boldsymbol{u}R(\boldsymbol{u}R)\tilde{R}\boldsymbol{u} = -\boldsymbol{u}R$ (here we have used geometric covariance). Therefore, we also have $(m_{\boldsymbol{u}}f)((\boldsymbol{u}R)_1) = -(\boldsymbol{u}R)_1$, and this implies that $m_{\boldsymbol{u}}f = m_{\boldsymbol{v}}$, where $\boldsymbol{v} = (\boldsymbol{u}R)_1$. Now, $\boldsymbol{u}R = \boldsymbol{u}\cos\frac{\alpha}{2} - \boldsymbol{u\omega i}\sin\frac{\alpha}{2}$, and so

$$\boldsymbol{v} = (\boldsymbol{u}R)_1 = \boldsymbol{u}\cos\tfrac{\alpha}{2} - (\boldsymbol{u}\wedge\boldsymbol{\omega})\boldsymbol{i}\sin\tfrac{\alpha}{2} = \boldsymbol{u}\cos\tfrac{\alpha}{2} + (\boldsymbol{u}\times\boldsymbol{\omega})\sin\tfrac{\alpha}{2}.$$

We have used that $\boldsymbol{u\omega i} = (\boldsymbol{u}\cdot\boldsymbol{\omega})\boldsymbol{i} + (\boldsymbol{u}\wedge\boldsymbol{\omega})\boldsymbol{i}$ and that its vector part is $(\boldsymbol{u}\wedge\boldsymbol{\omega})\boldsymbol{i}$.

5.1.10 ($\mathcal{R}_{2,1} \simeq \mathrm{SL}_2$) We can use E.3.6, p. 58 and E.3.4, p. 58 to get

$$\mathcal{G}_{2,1} \simeq \mathcal{G}_{1,0}(2) \simeq 2\mathbb{R}(2) = \mathbb{R}(2) \oplus \mathbb{R}(2). \qquad (5.2)$$

To describe the first isomorphism in more detail, let $\boldsymbol{e}_0, \boldsymbol{e}_1, \boldsymbol{e}_2$ be an orthonormal basis of $E_{2,1}$, and let $\boldsymbol{e}_1$ and $\boldsymbol{e}_2$ play the role of $\boldsymbol{e}$ and $\bar{\boldsymbol{e}}$ in E.3.6. Then, an even element of $\mathcal{G}_{2,1}$ can be written in a unique way in the form:

$$R = \alpha + (\beta_1\boldsymbol{e}_0)\boldsymbol{e}_1 + (\beta_2\boldsymbol{e}_0)\boldsymbol{e}_2 + \gamma\boldsymbol{e}_1\boldsymbol{e}_2, \quad \alpha, \beta_1, \beta_2, \gamma \in \mathbb{R}.$$

The matrix corresponding to R by the isomorphism in that exercise is

$$M = \begin{pmatrix} \alpha + \beta_1 \boldsymbol{e}_0 & \gamma + \beta_2 \boldsymbol{e}_0 \\ \gamma - \beta_2 \boldsymbol{e}_0 & \alpha - \beta_1 \boldsymbol{e}_0 \end{pmatrix} = \begin{pmatrix} z & w \\ \hat{w} & \hat{z} \end{pmatrix},$$

with $z = \alpha + \beta_1 \boldsymbol{e}_0$ and $w = \gamma + \beta_2 \boldsymbol{e}_0$, so that $z, w \in \mathcal{G}_1 = \langle 1, \boldsymbol{e}_0 \rangle$. Since

$$\tilde{R} = \alpha - (\beta_1 \boldsymbol{e}_0)\boldsymbol{e}_1 - (\beta_2 \boldsymbol{e}_0)\boldsymbol{e}_2 - \gamma \boldsymbol{e}_1 \boldsymbol{e}_2,$$

we can write

$$\tilde{R} \equiv \begin{pmatrix} \hat{z} & -w \\ -\hat{w} & z \end{pmatrix}$$

and it immediately follows that the rotor condition $R\tilde{R} = 1$ is equivalent to the relation $z\hat{z} - w\hat{w} = 1$, or $\det(M) = 1$.

If we want the actual decomposition of the matrix M into a pair of real matrices, we only have to remember, with the notations of E.3.4, that

$$M = Mu + Mu' = \begin{pmatrix} \alpha + \beta_1 & \gamma + \beta_2 \\ \gamma - \beta_2 & \alpha - \beta_1 \end{pmatrix} u + \begin{pmatrix} \alpha - \beta_1 & \gamma - \beta_2 \\ \alpha + \beta_2 & \gamma + \beta_1 \end{pmatrix} u'$$

and the condition that R is a rotor is expressed by $\det(Mu) = 1$ (or $\det(Mu') = 1$). The final conclusion is that the map $R \mapsto Mu$ yields an isomorphism:

$$\mathcal{R}_{2,1} \simeq \mathrm{SL}_2,$$

as any matrix of GL_2 can be written in a unique way in the form:

$$M' = \begin{pmatrix} \alpha + \beta_1 & \gamma + \beta_2 \\ \gamma - \beta_2 & \alpha - \beta_1 \end{pmatrix}, \quad \alpha, \beta_1, \beta_2, \gamma \in \mathbb{R},$$

and $M' \in \mathrm{SL}_2$ if and only if $1 = \det(M') = \alpha^2 + \beta_2^2 - (\gamma^2 + \beta_1^2) = \det(M)$, that is, if and only if $R \in \mathcal{R}_{2,1}$.

Remark that Eq. (5.2) can also be deduced on noting that the signature (2, 1) is special, and so we can use E.3.9 (6) to conclude that $\mathcal{G}_{2,1} \simeq 2\mathcal{G}_{2,1}^+$, and then E.3.8, p. 60, yields that it is $\simeq 2E_{1,1} \simeq 2\mathbb{R}(2)$.

5.1.11 ($\mathcal{R}_{1,2} \simeq \mathrm{SL}_2$) Since the even algebras $\mathcal{G}_{1,2}^+$ and $\mathcal{G}_{2,1}^+$ are both isomorphic to $\mathcal{G}_{1,1}$ (see E.3.8, p. 60), the group $\mathcal{R}_{1,2}$ is also isomorphic to SL_2. It is instructive, however, to determine $\mathcal{R}_{1,2}$ as in the previous case and then see how the isomorphism works.

We can use the isomorphism $\mathcal{G}_{1,2} \simeq \mathcal{G}_{0,1}(2) \simeq \mathbb{C}(2)$ established in E.3.6, p. 58. If $\boldsymbol{e}_0, \boldsymbol{e}_1, \boldsymbol{e}_2$ is an orthonormal basis of $E_{1,2}$, we may take, using the notations of that exercise, e_0, e_1 to play the role of e and $\bar{e}$ and $\langle 1, e_2\rangle = \mathcal{G}_{0,1}$ as $\mathbb{C}$. It follows that the even elements of $\mathcal{G}_{1,2}$ can be written in a unique way in the form:

$$R = \alpha + (\beta_1 e_2)e_0 + (\beta_2 e_2)e_1 + \gamma e_0 e_1, \quad \alpha, \beta_1, \beta_2, \gamma \in \mathbb{R},$$

and the matrix in $\mathcal{G}_{0,1}(2)$ corresponding to R is

$$M = \begin{pmatrix} \alpha + \beta_1 e_2 & \gamma + \beta_2 e_2 \\ \gamma - \beta_2 e_2 & \alpha - \beta_1 e_2 \end{pmatrix} = \begin{pmatrix} z & w \\ \bar{w} & \bar{z} \end{pmatrix}, \tag{5.3}$$

where we set $\alpha + \beta_1 e_2 \equiv \alpha + \beta_1 i = z$, $\gamma + \beta_2 e_2 \equiv \gamma + \beta_2 i = w$. Since

$$\tilde{R} = \alpha - (\beta_1 e_2)e_0 - (\beta_2 e_2)e_1 - \gamma e_0 e_1,$$

its matrix representation is

$$\begin{pmatrix} \bar{z} & -w \\ -\bar{w} & z \end{pmatrix}.$$

It follows that $R\tilde{R}$ is represented by $(z\bar{z} - w\bar{w})I_2 = \det(M)I_2$, and hence the condition for R to be a rotor is that $\det(M) = 1$. It is a condition like for $\mathcal{R}_{2,1}$, but with β_1 and β_2 interchanged. After this observation, it can be checked that the map:

$$\begin{pmatrix} \alpha + \beta_1 i & \gamma + \beta_2 i \\ \gamma - \beta_2 i & \alpha - \beta_1 i \end{pmatrix} \mapsto \begin{pmatrix} \alpha + \beta_2 & \gamma + \beta_1 \\ \gamma - \beta_1 & \alpha - \beta_2 \end{pmatrix}$$

is an isomorphism $\mathcal{R}_{1,2} \simeq \mathrm{SL}_2$. Actually, the left matrix is the conjugate (in $\mathrm{GL}_2(\mathbb{C})$) of the right matrix by the matrix $\begin{pmatrix} 1 & -i \\ 1 & i \end{pmatrix}$.

In the representation of $\mathcal{R}_{1,2}$ provided by Eq. (5.3), we have $|z|^2 - |w|^2 = 1$. So, $|z| \geqslant 1$ and $|w/z|^2 = 1 - 1/|z|^2 < 1$. Therefore, we have a map $\mathcal{R}_{1,2} \to U_1 \times \Delta$, where Δ denotes the open disc of radius 1, given by:

$$M(z, w) = \begin{pmatrix} z & w \\ \bar{w} & \bar{z} \end{pmatrix} \mapsto (z/|z|, w/z).$$

This map is a homeomorphism, with inverse the map

$$(u, v) \mapsto \rho M(u, vu), \quad \rho = 1/(1 - |v|^2)^{1/2}.$$

5.2 Plane Rotors

The most simple rotors are given as the product of two unit vectors of the same signature. We devote this section to them because they can be used, as we will see, to construct all rotors, and hence all pinors and versors.

Given two linearly independent unit vectors $\boldsymbol{u}$ and $\boldsymbol{v}$, $R = \boldsymbol{v}\boldsymbol{u}$ is a spinor. We will say that it is the *plane spinor* defined by $\boldsymbol{u}$ and $\boldsymbol{v}$. Since $R\tilde{R} = \boldsymbol{u}^2\boldsymbol{v}^2$, we see that R is a rotor (in which case we will say that it is a *plane rotor*) if and only if $\boldsymbol{u}^2\boldsymbol{v}^2 = 1$, that is, if either $\boldsymbol{u}^2 = \boldsymbol{v}^2 = 1$ or else $\boldsymbol{u}^2 = \boldsymbol{v}^2 = -1$. In both cases:

$$R = \boldsymbol{u} \cdot \boldsymbol{v} - \boldsymbol{u} \wedge \boldsymbol{v}$$

and

$$(\boldsymbol{u} \wedge \boldsymbol{v})^2 = (\boldsymbol{u} \wedge \boldsymbol{v}) \cdot (\boldsymbol{u} \wedge \boldsymbol{v}) = (\boldsymbol{u} \cdot \boldsymbol{v})^2 - \boldsymbol{u}^2\boldsymbol{v}^2 = (\boldsymbol{u} \cdot \boldsymbol{v})^2 - 1.$$

Thus, plane rotors are naturally classified according to the sign of this quantity. Notice that by 3.3.11 we have that the plane $|\boldsymbol{u} \wedge \boldsymbol{v}\rangle = \langle \boldsymbol{u}, \boldsymbol{v}\rangle$ is singular if and only if $(\boldsymbol{u} \wedge \boldsymbol{v})^2 = 0$.

Elliptic Plane Rotors This is the case $(\boldsymbol{u} \wedge \boldsymbol{v})^2 < 0$, or $(\boldsymbol{u} \cdot \boldsymbol{v})^2 < 1$. Thus, there exists a unique $\alpha \in (0, \pi)$ such that $\boldsymbol{u} \cdot \boldsymbol{v} = \cos\alpha$ and

$$(\boldsymbol{u} \wedge \boldsymbol{v})^2 = \cos^2\alpha - 1 = -\sin^2\alpha.$$

If we set $U = (\boldsymbol{u} \wedge \boldsymbol{v})/\sin\alpha$, then we have that $U^2 = -1$ and hence

$$R = \cos\alpha - U\sin\alpha = e^{-U\alpha}. \tag{5.4}$$

In this case, the regular plane $\langle \boldsymbol{u}, \boldsymbol{v}\rangle$ does not have non-zero isotropic vectors (check that the discriminant of the quadratic $(\lambda\boldsymbol{u} + \boldsymbol{v})^2 = 0$ is $4(\boldsymbol{u} \wedge \boldsymbol{v})^2$) and hence it is Euclidean if $\boldsymbol{u}^2 = \boldsymbol{v}^2 = 1$ and anti-Euclidean if $\boldsymbol{u}^2 = \boldsymbol{v}^2 = -1$.

Hyperbolic Plane Rotors If $(\boldsymbol{u} \wedge \boldsymbol{v})^2 > 0$, then $(\boldsymbol{u} \cdot \boldsymbol{v})^2 > 1$, and there exists a unique $\alpha > 0$ such that $\boldsymbol{u} \cdot \boldsymbol{v} = \epsilon\cosh\alpha$, where now ϵ is the sign of $\boldsymbol{u} \cdot \boldsymbol{v}$, and

$$(\boldsymbol{u} \wedge \boldsymbol{v})^2 = \cosh^2\alpha - 1 = \sinh^2\alpha.$$

Setting $U = (\boldsymbol{u} \wedge \boldsymbol{v})/\sinh\alpha$, we have that $U^2 = 1$ and hence

$$R = \epsilon\cosh\alpha - U\sinh\alpha = \epsilon e^{-U\epsilon\alpha}. \tag{5.5}$$

In this case, the regular plane $\langle \boldsymbol{u}, \boldsymbol{v}\rangle$ has non-zero isotropic vectors (the discriminant of the quadratic $(\lambda \boldsymbol{u} + \boldsymbol{v})^2 = 0$ is positive in this case), so that it is a hyperbolic plane (this means that it has signature (1,1)). Actually, it is easy to see that $\boldsymbol{w}^{\pm} = \epsilon e^{\pm\alpha}\boldsymbol{u} - \boldsymbol{v}$ are isotropic (see E.5.7, p. 104).

Parabolic Plane Rotors If $(\boldsymbol{u} \wedge \boldsymbol{v})^2 = 0$, then $(\boldsymbol{u} \cdot \boldsymbol{v})^2 = 1$, or $\boldsymbol{u} \cdot \boldsymbol{v} = \epsilon = \pm 1$, and so

$$R = \epsilon(1 - \epsilon \boldsymbol{u} \wedge \boldsymbol{v}) = \epsilon e^{-\epsilon \boldsymbol{u}\wedge\boldsymbol{v}}. \tag{5.6}$$

If this is the case, we know that the plane $\langle \boldsymbol{u}, \boldsymbol{v}\rangle$ is singular. This can be checked directly on noticing that the vector $\epsilon \boldsymbol{v} - \epsilon_{\boldsymbol{u}} \boldsymbol{u}$ is orthogonal to $\boldsymbol{u}$ and $\boldsymbol{v}$.

5.2.1 Up to sign, all plane rotors are exponentials of bivectors, and they are connected to 1 or to -1 (by this we mean that there is a continuous path on the rotor group joining the given rotor to either 1 or -1). Indeed, in the parabolic case $R = \epsilon e^{-\epsilon \boldsymbol{u}\wedge\boldsymbol{v}}$ is connected to ϵ by the rotor path:

$$R(t) = \epsilon e^{-\epsilon t \boldsymbol{u}\wedge\boldsymbol{v}} \quad (0 \leqslant t \leqslant 1),$$

as $R(1) = R$ and $R(0) = \epsilon$.

Similarly, any elliptic plane rotor is connected to 1 (it is enough to let the angle α in Eq. (5.4) vary in the interval $[0, 1]$) and any hyperbolic rotor (5.5) is connected to 1 if $\boldsymbol{u} \cdot \boldsymbol{v} > 0$ and to -1 if $\boldsymbol{u} \cdot \boldsymbol{v} < 0$ (let the parameter α in Eq. (5.5) vary in the interval $[0, 1]$).

5.2.2 ($\mathcal{R}$ is Generated by Plane Rotors)

(1) *Any rotor can be expressed as a product of plane rotors.*
(2) *Any rotor is connected to* 1 *or* -1.
(3) *If* $r \geqslant 2$ *or* $s \geqslant 2$, *then any rotor is connected to* 1. *Consequently,* $\mathcal{R}$ *is connected.*

Proof

(1) We will use that if v is a versor and $\boldsymbol{u}$ a unit vector, then $v\boldsymbol{u} = \boldsymbol{u}'v$, with $\boldsymbol{u}'$ a vector that has the same signature as $\boldsymbol{u}$. Indeed, we can write

$$v\boldsymbol{u} = v\boldsymbol{u}v^{-1}v = \pi(v)\,\underline{v}(\boldsymbol{u})v$$

and $\boldsymbol{u}' = \pi(v)\,\underline{v}(\boldsymbol{u})$ satisfies that $q(\boldsymbol{u}') = q(\boldsymbol{u})$ because $\underline{v}$ is an isometry.

Now, let $R \in \mathcal{R}$, say $R = \boldsymbol{u}_1 \cdots \boldsymbol{u}_{2k}$, with the $\boldsymbol{u}_j$ unit vectors. Using the preceding observation, we can express R as a product of unit vectors in such a way that no positive vector appears later than a negative one. In fact, if a positive vector $\boldsymbol{u}$ in R appears just after a stretch of negative vectors v in R, we can replace $v\boldsymbol{u}$ by $\boldsymbol{u}'v$, with $\boldsymbol{u}'$ positive. This replacement can go on until there is no positive vector that appears to the right of a negative one.

Since R is a rotor, the number of negative vectors in it must be even, and consequently the number of positive vectors is also even. Grouping the vectors in consecutive pairs, we get that R is the product of k plane rotors.

(2) Since each of the k plane rotors is connected to 1 or to -1, the same is true for their product. So, R can be connected to 1 or to -1. Actually, if the number of plane rotors connected to -1 is even (odd), then R is connected to 1 (to -1).

(3) If $r \geqslant 2$ or $s \geqslant 2$, we can choose two unit vectors of the same signature, say $\boldsymbol{e}_1$ and $\boldsymbol{e}_2$. Then, $(\boldsymbol{e}_1\boldsymbol{e}_2)^2 = -1$, and the rotor path:

$$t \mapsto e^{t\boldsymbol{e}_1\boldsymbol{e}_2} = \cos t + \boldsymbol{e}_1\boldsymbol{e}_2 \sin t \quad (0 \leqslant t \leqslant \pi)$$

connects 1 ($t = 0$) to -1 ($t = \pi$). □

5.2.3 *If* $r \geqslant 2$ *or* $s \geqslant 2$, SO^0 *is the connected component subgroup of* SO.

Proof Since $\mathcal{R}$ is connected, its image SO^0 in SO is also connected. If $s = 0$ or $r = 0$, $\mathrm{SO} = \mathrm{SO}^0$ and so SO is connected. Otherwise, SO is the image of $\mathcal{S}$, and hence it is the disjoint union of the image SO^0 of $\mathcal{R}$ and, using notations of 4.3.4, the image $\rho(\boldsymbol{u}\bar{\boldsymbol{u}})\mathrm{SO}^0$ of $\boldsymbol{u}\bar{\boldsymbol{u}}\mathcal{R}$. Note that if we take $\boldsymbol{u} = \boldsymbol{e}_1$ and $\bar{\boldsymbol{u}} = \boldsymbol{e}_n$, then $\rho(\boldsymbol{e}_1\boldsymbol{e}_2)$ is the isometry that leaves $\boldsymbol{e}_1$ and $\boldsymbol{e}_n$ invariant and such that $\boldsymbol{e}_j \mapsto -\boldsymbol{e}_j$ for $j \neq 1, n$. □

5.3 Infinitesimal Rotations

Given a differentiable path $f : (-\tau, \tau) \to \mathrm{O}_{r,s}$ ($\tau > 0$) such that $f(0) = \mathrm{Id}$, we can take the derivative $h = \frac{df}{dt}\big|_{t=0}$. Since the vector space that contains $\mathrm{O}_{r,s}$ is $\mathrm{End}(E)$, f is also a path on this space and hence $h \in \mathrm{End}(E)$. Such endomorphisms h are said to be *infinitesimal rotations*. Since for any fixed vectors $\boldsymbol{x}$ and $\boldsymbol{y}$, we have that

$$(f(t)\boldsymbol{x}) \cdot (f(t)\boldsymbol{y}) = \boldsymbol{x} \cdot \boldsymbol{y},$$

taking the derivative with respect to t at $t = 0$ we get

$$(h\boldsymbol{x}) \cdot \boldsymbol{y} + \boldsymbol{x} \cdot (h\boldsymbol{y}) = 0,$$

so that *any infinitesimal rotation is a skew-symmetric endomorphism* (in 5.3.3 we provide more details about the definition of these endomorphisms).

The converse is also true. To see this, notice that to be skew-symmetric is equivalent to the relation $(h\boldsymbol{x}) \cdot \boldsymbol{y} = \boldsymbol{x} \cdot (-h\boldsymbol{y})$ for any vectors $\boldsymbol{x}$ and $\boldsymbol{y}$. This implies that

$$(h^m\boldsymbol{x}) \cdot \boldsymbol{y} = \boldsymbol{x} \cdot \big((-h)^m \boldsymbol{y}\big),$$

for any positive integer m, and hence that

$$\big(P(h)\boldsymbol{x}\big) \cdot \boldsymbol{y} = \boldsymbol{x} \cdot \big(P(-h)\boldsymbol{y}\big)$$

for any univariate polynomial P. If we take for P a partial sum P_m of the exponential function e^x up to degree m, so that $P_m(h) = \sum_{j=0}^{m} h^j/j!$, and define

$$e^h = \lim_{m\to\infty} P_m(h)$$

(note that this limit is well defined in the space $\mathrm{End}(E)$), we conclude that we also have

$$\big(e^h\boldsymbol{x}\big) \cdot \boldsymbol{y} = \boldsymbol{x} \cdot \big(e^{-h}\boldsymbol{y}\big).$$

Replacing $\boldsymbol{y}$ by $e^h\boldsymbol{y}$ yields

$$\big(e^h\boldsymbol{x}\big) \cdot \big(e^h\boldsymbol{y}\big) = \boldsymbol{x} \cdot \boldsymbol{y},$$

which means that e^h is an isometry. If we now take $f(t) = e^{th}$, this is a differentiable path on $\mathrm{O}_{r,s}$, defined for all $t \in \mathbb{R}$, and clearly $\frac{df}{dt}\big|_{t=0} = h$. Thus, we have the following result.

5.3.1 *An endomorphism h of (E, q) is an infinitesimal rotation if and only if it is skew-symmetric.*

5.3.2 (Example: The Infinitesimal Rotation About an Axis of E_3) If we let $\boldsymbol{u}$ be any non-zero vector of the Euclidean space E_3, we can use Eq. (4.7) to find the infinitesimal rotation associated to the path $f_{t\boldsymbol{u}} \in \mathrm{SO}_3$ ($t \in \mathbb{R}$). Note that the expression:

$$f_{t\boldsymbol{u}}(\boldsymbol{x}) = e^{-\mathbf{i}\boldsymbol{u}t/2}\boldsymbol{x}e^{\mathbf{i}\boldsymbol{u}t/2}$$

insures that $f_{t\boldsymbol{u}}$ is differentiable with respect to t and that $f_{\mathbf{0}} = \mathrm{Id}$, so that the conditions needed for getting an infinitesimal rotation are fulfilled. If we let $r_{\boldsymbol{u}}$ denote the associated infinitesimal rotation, we get, for an arbitrary fixed vector $\boldsymbol{x}$,

$$r_{\boldsymbol{u}}(\boldsymbol{x}) = \boldsymbol{u} \times \boldsymbol{x},$$

where $\boldsymbol{u} \times \boldsymbol{x}$ is the cross product of $\boldsymbol{u}$ and $\boldsymbol{x}$. Indeed,

$$\begin{aligned} r_{\boldsymbol{u}}(\boldsymbol{x}) &= \tfrac{d}{dt}\big(e^{-\mathbf{i}\boldsymbol{u}t/2}\boldsymbol{x}e^{\mathbf{i}\boldsymbol{u}t/2}\big)\big|_{t=0} \\ &= -\tfrac{1}{2}\mathbf{i}\boldsymbol{u}\boldsymbol{x} + \tfrac{1}{2}\boldsymbol{x}\mathbf{i}\boldsymbol{u} \\ &= -\tfrac{1}{2}(\boldsymbol{u}\boldsymbol{x} - \boldsymbol{x}\boldsymbol{u})\mathbf{i} \end{aligned}$$

$$= -(\boldsymbol{u} \wedge \boldsymbol{x})\mathbf{i}$$
$$= \boldsymbol{u} \times \boldsymbol{x}.$$

We have used that $\mathbf{i}$ commutes with vectors. The last step is nothing but the definition of the cross product as the Hodge dual of the wedge product $\boldsymbol{u} \wedge \boldsymbol{x}$ (for a systematic geometric algebra account of the properties of the cross product, we may refer to [98]). To note that in this case we can check directly that $r_{\boldsymbol{u}}$ is skew-symmetric. Indeed,

$$(r_{\boldsymbol{u}}\boldsymbol{x}) \cdot \boldsymbol{x} = (\boldsymbol{u} \times \boldsymbol{x}) \cdot \boldsymbol{x} = 0,$$

as $\boldsymbol{u} \times \boldsymbol{x}$ is orthogonal to $\boldsymbol{x}$.

This example also works for the Euclidean space E_n ($n \geqslant 3$), but using the general spinorial formula (4.6′). That formula supplies the rotation by an angle α in the plane of the area element $\boldsymbol{a} = \boldsymbol{i}\alpha$ (where $\boldsymbol{i}$ is the unit area of that plane giving the positive sense of the rotation). We can therefore regard $f_{\boldsymbol{a},t} = f_{\boldsymbol{a}t}$ ($t \in \mathbb{R}$) as a differentiable path on SO_n. Since $f_{\boldsymbol{a},0} = \mathrm{Id}$, this path defines an infinitesimal rotation $r_{\boldsymbol{a}}$:

$$r_{\boldsymbol{a}}(\boldsymbol{x}) = \tfrac{d}{dt}\left(e^{-\boldsymbol{a}t/2}\boldsymbol{x}e^{\boldsymbol{a}t/2}\right)\Big|_{t=0} = -\tfrac{1}{2}(\boldsymbol{a}\boldsymbol{x} - \boldsymbol{x}\boldsymbol{a}) = \boldsymbol{x} \cdot \boldsymbol{a},$$

where in the last step we use Riesz's second formula 3.3.6 (note that $\hat{\boldsymbol{a}} = \boldsymbol{a}$). The fact that $r_{\boldsymbol{a}}$ is skew-symmetric can also be checked directly:

$$\boldsymbol{x} \cdot (r_{\boldsymbol{a}}\boldsymbol{x}) = \boldsymbol{x} \cdot (\boldsymbol{x} \cdot \boldsymbol{a}) = (\boldsymbol{x} \wedge \boldsymbol{x}) \cdot \boldsymbol{a} = 0.$$

In the case of dimension 3, the infinitesimal rotations $r_{\boldsymbol{u}^*}$ and $r_{\boldsymbol{u}}$ are the same, as expected by consistency, for

$$r_{\boldsymbol{u}^*}(\boldsymbol{x}) = \tfrac{1}{2}(\boldsymbol{x}\boldsymbol{u}^* - \boldsymbol{u}^*\boldsymbol{x}) = \tfrac{1}{2}(\boldsymbol{x}\boldsymbol{u}\mathbf{i} - \boldsymbol{u}\mathbf{i}\boldsymbol{x}) = \tfrac{1}{2}(\boldsymbol{x}\boldsymbol{u} - \boldsymbol{u}\boldsymbol{x})\mathbf{i}$$
$$= (\boldsymbol{x} \wedge \boldsymbol{u})\mathbf{i} = -\boldsymbol{x} \times \boldsymbol{u} = \boldsymbol{u} \times \boldsymbol{x} = r_{\boldsymbol{u}}(\boldsymbol{x}).$$

The case of the Euclidean plane E_2 is also very instructive, but we leave it as an exercise (see E.5.8, p. 104).

5.3.3 (Background Remarks on the Adjoint of an Endomorphism) Given an endomorphism $f : E \to E$, there exists a unique endomorphism $f^\dagger : E \to E$, called the *q-adjoint* of f, or simply the *adjoint* when q can be understood, such that

$$(f\boldsymbol{x}) \cdot \boldsymbol{y} = \boldsymbol{x} \cdot (f^\dagger \boldsymbol{y}) \text{ for all } \boldsymbol{x}, \boldsymbol{y} \in E. \tag{5.7}$$

Moreover, the map $f \mapsto f^\dagger$ is linear and satisfies

$$(gf)^\dagger = f^\dagger g^\dagger$$

for all $f, g \in \mathrm{End}(E)$ (see E.5.9, p. 104).

Thus, an endomorphism h is skew-symmetric if and only if $h^{\dagger} = -h$. The endomorphisms f that satisfy $f^{\dagger} = f$ are called *symmetric*. Note also that if f is an *orthogonal endomorphism* (that is, an isometry), then $f^{\dagger} = f^{-1}$. Indeed, by definition f satisfies $(f\boldsymbol{x}) \cdot (f\boldsymbol{y}) = \boldsymbol{x} \cdot \boldsymbol{y}$ for all vectors $\boldsymbol{x}$ and $\boldsymbol{y}$. Since q is regular, this means that f is an isometry. Replacing $\boldsymbol{y}$ by $f^{-1}\boldsymbol{y}$, we get $(f\boldsymbol{x})\cdot\boldsymbol{y} = \boldsymbol{x}\cdot(f^{-1}\boldsymbol{y})$, and this proves the claim.

Symmetric, skew-symmetric, and orthogonal endomorphisms are special cases of what are called *normal endomorphisms*, which by definition are those that commute with their adjoint: $ff^{\dagger} = f^{\dagger}f$. As a prerequisite for our considerations about the exponential maps, we include a subsection in Sect. 5.4 to provide a brief account about the structure of such endomorphisms in the case of the Euclidean space E_n.

5.3.4 (The Lie Algebra $\mathfrak{so}(E, q)$) The skew-symmetric endomorphisms of (E, q) form a vector subspace of $\mathrm{End}(E)$. This space is closed under the *commutator bracket*, $[h, h'] = hh' - h'h$, because

$$[h, h']^{\dagger} = h'^{\dagger}h^{\dagger} - h^{\dagger}h'^{\dagger} = (-h')(-h) - (-h)(-h') = h'h - hh' = -[h, h'].$$

We shall write $\mathfrak{so}(E, q)$, or just $\mathfrak{so}(E)$, to denote the space of skew-symmetric endomorphisms of (E, q) endowed with the commutator bracket. This bracket is *bilinear, skew-symmetric* (meaning that $[h, h'] = -[h', h]$ for all $h, h' \in \mathfrak{so}(E, q)$, or, equivalently, that $[h, h] = 0$ for all $h \in \mathfrak{so}(E)$), and it is easy to check that it satisfies the *Jacobi identity* (see E.5.10, p. 105):

$$[h, [h', h'']] + [h', [h'', h]] + [h'', [h, h']] = 0.$$

These properties are the defining axioms for a Lie algebra, and so it is said that $\mathfrak{so}(E)$ is the *Lie algebra* of $\mathrm{SO}(E)$, and it is denoted $\mathfrak{so}_{r,s}$ when we want to specify the signature. Let us also note that by E.5.9, p. 104, we have

$$\dim \mathfrak{so}(E) = \binom{n}{2}. \tag{5.8}$$

5.3.5 (Example: $[r_{\boldsymbol{u}}, r_{\boldsymbol{v}}]$) In Example 5.3.2, we have found that for any vector $\boldsymbol{u} \in E_3$ we have an infinitesimal rotor $r_{\boldsymbol{u}} \in \mathfrak{so}_3$ defined by $r_{\boldsymbol{u}}(\boldsymbol{x}) = \boldsymbol{u} \times \boldsymbol{x}$. If $\boldsymbol{v}$ is another vector, then we have

$$[r_{\boldsymbol{u}}, r_{\boldsymbol{v}}] = r_{\boldsymbol{u}\times\boldsymbol{v}}.$$

This follows from the double cross product formula (cf. [98, 1.3.15 (4)]):

$$\begin{aligned}(r_{\boldsymbol{u}}r_{\boldsymbol{v}} - r_{\boldsymbol{v}}r_{\boldsymbol{u}})(\boldsymbol{x}) &= \boldsymbol{u} \times (\boldsymbol{v} \times \boldsymbol{x}) - \boldsymbol{v} \times (\boldsymbol{u} \times \boldsymbol{x}) \\ &= (\boldsymbol{u} \cdot \boldsymbol{x})\boldsymbol{v} - (\boldsymbol{u} \cdot \boldsymbol{v})\boldsymbol{x} - (\boldsymbol{v} \cdot \boldsymbol{x})\boldsymbol{u} + (\boldsymbol{v} \cdot \boldsymbol{u})\boldsymbol{x} \\ &= (\boldsymbol{u} \cdot \boldsymbol{x})\boldsymbol{v} - (\boldsymbol{v} \cdot \boldsymbol{x})\boldsymbol{u}\end{aligned}$$

$$= (\boldsymbol{u} \times \boldsymbol{v}) \times \boldsymbol{x}$$
$$= r_{\boldsymbol{u}\times\boldsymbol{v}}(\boldsymbol{x}).$$

Actually, the double cross product formula implies that $(E_3, \times)$ is a Lie algebra, and the present example shows that the map $E_3 \to \mathfrak{so}_3$, $\boldsymbol{u} \mapsto r_{\boldsymbol{u}}$, is a homomorphism of Lie algebras. This homomorphism is one-to-one, because $\boldsymbol{u} \times \boldsymbol{x} = \boldsymbol{0}$ for all vectors $\boldsymbol{x}$ can happen only for $\boldsymbol{u} = \boldsymbol{0}$. Since by Eq. (5.8) we have that $\dim \mathfrak{so}_3 = 3$, we conclude that $(E_3, \times) \simeq \mathfrak{so}_3$.

5.4 Infinitesimal Rotors

In the previous section, we have used some of the goods provided by geometric algebra, but we have been working at the level of the orthogonal and special orthogonal groups. We know, however, that the management of such groups with geometric algebra is mediated by the rotor and spinor groups. Thus, we should expect to reap a richer harvest if we succeeded in developing a working scheme for the rotor group similar to the one followed for $\mathrm{SO}(E)$ and in elucidating the relations between the two views.

The first idea is to introduce the notion of infinitesimal rotor by analogy to the notion of infinitesimal rotation. Since the rotor group $\mathcal{R} = \mathcal{R}_{r,s}$ is part of the even algebra $\mathcal{G}^+ = \mathcal{G}^+_{r,s}$, this should be feasible by using the linear structure of this space to take derivatives of paths traced on $\mathcal{R}$, so that in principle an infinitesimal rotor will be some kind of even multivector.

In more concrete terms, if we have a differentiable rotor path:

$$R : (-\tau, \tau) \to \mathcal{R} \quad (\tau > 0)$$

such that $R(0) = 1$, then we have

$$R'(0) = \tfrac{dR(t)}{dt}\Big|_{t=0} \in \mathcal{G}^+$$

and the even multivectors got in this way are called *infinitesimal rotors*.

5.4.1 (Examples: Plane Rotor Paths) Consider an elliptic rotor $R(\alpha) = e^{-U\alpha}$, where U is a bivector such that $U^2 = -1$ (see Eq. (5.4)). Since R is differentiable and $R(0) = 1$, we get that $R'(0) = -U$ is an infinitesimal rotor. The same happens with the hyperbolic rotor $R(\alpha) = \epsilon e^{-U\epsilon\alpha}$, where U is a bivector such that $U^2 = 1$ (see Eq. (5.5)), that tells us that $-U$ is an infinitesimal rotor (the sign ϵ disappears when taking the derivative). A similar reasoning allows us to see that $-\boldsymbol{u} \wedge \boldsymbol{v}$ is an infinitesimal rotor if $\epsilon e^{-\epsilon \boldsymbol{u}\wedge\boldsymbol{v}}$ is a parabolic rotor (see Eq. (5.6)). Note that all these infinitesimal rotors are bivectors. Next result, whose proof is adapted from [106, Section 6.5], shows that this is not an accident.

5.4.2 *Any infinitesimal rotor is a bivector.*

Proof Taking the derivative of the expression $R(t)\widetilde{R(t)} = 1$ with respect to t at 0, we obtain that $\widetilde{R'(0)} = -R'(0)$. Since $R'(0)$ is clearly an even multivector, the last condition shows that the grades $k = 2j$ of $R'(0)$ such that $R'(0)_k \neq 0$ must satisfy that $k/\!\!/2 = j$ has to be odd, say $j = 2l + 1$, which implies that $k = 4l + 2$, $l \geqslant 0$. In particular, we get that $R'(0) = b + z$, where b is a bivector and z a multivector whose least grade is $\geqslant 6$.

Now, let us use that $\boldsymbol{x}(t) = R(t)\,\boldsymbol{x}\,\widetilde{R(t)} \in E$ for any given $\boldsymbol{x} \in E$. Taking the derivative of this relation with respect to t at 0, we have (for the last step, use Riesz formulas 3.3.6):

$$\begin{aligned}\boldsymbol{x}'(0) &= R'(0)\boldsymbol{x} + \boldsymbol{x}\widetilde{R'(0)} = (b+z)\boldsymbol{x} - \boldsymbol{x}(b+z)\\ &= b\boldsymbol{x} - \boldsymbol{x}b + z\boldsymbol{x} - \boldsymbol{x}z = 2b\cdot\boldsymbol{x} + 2z\cdot\boldsymbol{x}.\end{aligned}$$

Since $\boldsymbol{x}'(0)$ and $b \cdot \boldsymbol{x}$ are vectors and that the minimum grade of $z \cdot \boldsymbol{x}$ is $\geqslant 5$, we conclude that $z \cdot \boldsymbol{x} = 0$. Given that $\boldsymbol{x}$ is an arbitrary vector, 2.4.11 allows us to conclude that $z = 0$, and hence $R'(0) = b$ is a bivector. □

The converse of 5.4.2 is also true, as we will see in 5.4.7, but for the proof we need to take a short detour to establish some preliminary results which have independent interest. The key result will be 5.4.6, which gives a sense to the exponential of a bivector and shows that the result is a rotor. The purpose of statements 5.4.3 and 5.4.4 is to identify the bivector Lie algebra $\mathcal{G}^2$ (with the commutator bracket) with the Lie algebra of skew-symmetric endomorphisms of E (also with the commutator bracket).

The Lie Algebra Structure of $\mathcal{G}^2$

Given a bivector $b \in \mathcal{G}^2$, define the linear map $\mathsf{ad}_b : \mathcal{G} \to \mathcal{G}$ by the formula:

$$\mathsf{ad}_b(x) = [b, x] = bx - xb, \tag{5.9}$$

called the *commutator bracket* of b and x.

5.4.3 (Properties of ad_b) *For any bivector b, the operator ad_b preserves grades, and it is a derivation of the geometric product.*

Proof It is clear that ad_b vanishes on scalars. Since by the Riesz formulas 3.3.6 we have $\mathsf{ad}_b(\boldsymbol{x}) = 2b\cdot\boldsymbol{x} \in E$ for any vector $\boldsymbol{x}$, we also have that ad_b transforms vectors to vectors. Now, let $x \in \mathcal{G}^k$, $k \geqslant 2$. Then, 3.3.8 allows us to write

$$bx = b\cdot x + (bx)_k + b\wedge x \quad \text{and} \quad xb = x\cdot b + (xb)_k + x\wedge b.$$

But in this case, $x \cdot b = b \cdot x$ and $x \wedge b = b \wedge x$, and consequently $[b, x] = (bx)_k - (xb)_k$, which is a k-vector.

That ad_b is a derivation can be established with a simple computation:

$$\begin{aligned}\mathrm{ad}_b(xy) &= bxy - xyb \\ &= bxy - xby + xby - xyb \\ &= (bx - xb)y + x(by - yb) \\ &= \mathrm{ad}_b(x)\, y + x\, \mathrm{ad}_b(y).\end{aligned}$$

□

5.4.4

(1) *If* $b \in \mathcal{G}^2$, *then* $\mathrm{ad}_b \in \mathfrak{so}(E)$.
(2) *The map* $\mathcal{G}^2 \to \mathfrak{so}(E)$, $b \mapsto \mathrm{ad}_b$, *is a linear isomorphism.*

Proof

(1) The claim is clear from computation of $\mathrm{ad}_b(\boldsymbol{x}) \cdot \boldsymbol{y}$ and $\boldsymbol{x} \cdot \mathrm{ad}_b(\boldsymbol{y})$:

$$\mathrm{ad}_b(\boldsymbol{x}) \cdot \boldsymbol{y} = (b\boldsymbol{x} - \boldsymbol{x}b) \cdot \boldsymbol{y} = 2(b \cdot \boldsymbol{x}) \cdot \boldsymbol{y} = 2b \cdot (\boldsymbol{x} \wedge \boldsymbol{y}),$$
$$\boldsymbol{x} \cdot \mathrm{ad}_b(\boldsymbol{y}) = \boldsymbol{x} \cdot (b\boldsymbol{y} - \boldsymbol{y}b) = 2\boldsymbol{x} \cdot (b \cdot \boldsymbol{y}) = 2(b \cdot \boldsymbol{y}) \cdot \boldsymbol{x} = 2b \cdot (\boldsymbol{y} \wedge \boldsymbol{x}).$$

(2) The linear map $b \mapsto \mathrm{ad}_b$ is one-to-one. Indeed, if $\mathrm{ad}_b = 0$, then $\boldsymbol{x} \cdot b = 0$ for any vector $\boldsymbol{x}$ and this implies, by 2.4.11, that $b = 0$. On the other hand, we have $\dim \mathcal{G}^2 = \binom{n}{2} = \dim \mathfrak{so}(E)$, and therefore the map is also onto. □

5.4.5 (Remark: Inverse Isomorphism) If $f \in \mathfrak{so}(E)$, then $f = \mathrm{ad}_b$, with

$$b = \tfrac{1}{4} \sum_k f(\boldsymbol{e}_k) \wedge \boldsymbol{e}^k,$$

where $\boldsymbol{e}_1, \ldots, \boldsymbol{e}_n \in E$ is a basis, and $\boldsymbol{e}^1, \ldots, \boldsymbol{e}^n$ is the reciprocal basis (defined by the relations $\boldsymbol{e}^j \cdot \boldsymbol{e}_k = \delta^j_k$).

It suffices to see that $\mathrm{ad}_b(\boldsymbol{e}_j) = f(\boldsymbol{e}_j)$ for $j = 1, \ldots, n$. Indeed,

$$\mathrm{ad}_b(\boldsymbol{e}_j) = \tfrac{1}{2}\textstyle\sum_k (\boldsymbol{e}^k \cdot \boldsymbol{e}_j) f(\boldsymbol{e}_k) - \tfrac{1}{2}\sum_k (f(\boldsymbol{e}_k) \cdot \boldsymbol{e}_j)\boldsymbol{e}^k.$$

It is clear that the first term is $\frac{1}{2} f(\boldsymbol{e}_j)$. As for the second summand, it is equal to $\frac{1}{2} \sum_k (\boldsymbol{e}_k \cdot f(\boldsymbol{e}_j))\boldsymbol{e}^k$ (for f is skew-symmetric) and if we set $f(\boldsymbol{e}_j) = \sum_l a_{jl} \boldsymbol{e}_l$, then

$$\textstyle\sum_k (\boldsymbol{e}_k \cdot f(\boldsymbol{e}_j))\boldsymbol{e}^k = \sum_k \big(\sum_l a_{jl} g_{kl}\big)\boldsymbol{e}^k = \sum_l a_{jl} \sum_k g_{kl} \boldsymbol{e}^k = f(\boldsymbol{e}_j),$$

where $g_{kl} = \boldsymbol{e}_k \cdot \boldsymbol{e}_l$, and hence $\sum_k g_{kl} \boldsymbol{e}^k = \boldsymbol{e}_l$. □

The exponential e^x of any multivector x can be defined with the usual power series:

$$e^x = \sum_{k\geqslant 0} \frac{x^k}{k!}.$$

For its convergence, we use the linear structure of the geometric algebra $\mathcal{G}$. Note that the multivector e^x is even if x is even.

Next result, whose proof is inspired by Lundholm and Svensson [106, Theorem 6.17], shows that the exponential of a bivector is a rotor and provides a nice expression for the associated isometry.

5.4.6 *Let $b \in \mathcal{G}^2$, $\boldsymbol{x} \in E$, and $t \in \mathbb{R}$. Then:*

$$e^{tb}\boldsymbol{x}e^{-tb} = e^{t\,\mathsf{ad}_b}(\boldsymbol{x}) \quad \textit{and} \quad \pm e^b \in \mathcal{R}.$$

Proof The second relation is an easy consequence of the first. Indeed, $R = \pm e^b$ is an even multivector, it satisfies the condition $R\tilde{R} = 1$ because

$$\tilde{R} = \pm e^{\tilde{b}} = \pm e^{-b} = R^{-1},$$

and $R\boldsymbol{x}\tilde{R} = e^b\boldsymbol{x}e^{-b} = e^{\mathsf{ad}_b}(\boldsymbol{x}) \in E.$

To prove the first formula, consider the expression $f(t) = e^{tb}\boldsymbol{x}e^{-tb} \in \mathcal{G}$. Then, f is a differentiable function and

$$f'(t) = bf(t) - f(t)b = \mathsf{ad}_b(f(t)).$$

Iterating, and using that ad_b is a linear function independent of t, we get

$$f^{(k)}(t) = \frac{d^k}{dt^k} f(t) = \mathsf{ad}_b^k(f(t)).$$

In particular, $f^{(k)}(0) = \mathsf{ad}_b^k(\boldsymbol{x})$ for all k. This implies that the m-th Taylor polynomial of $f(t)$ at $t = 0$, namely $T_m f(t) = \sum_{k=1}^m \frac{1}{k!} f^{(k)}(0)t^k$, is given by:

$$T_m f(t) = \left(\sum_{k=1}^m \frac{1}{k!}\mathsf{ad}_b^k t^k\right)(\boldsymbol{x}).$$

Taking the limit for $m \to \infty$, the right-hand side converges to $e^{t\,\mathsf{ad}_b}(\boldsymbol{x})$ and the left-hand side converges to $f(t)$ because this function is actually *analytic* (see E.5.12, p. 106), and this completes the proof. □

5.4.7 *If $b \in \mathcal{G}^2$, then b is an infinitesimal rotor.*

Proof By the preceding result, $R(t) = e^{tb} \in \mathcal{R}$ for all t, and clearly $R'(0) = b$. □

5.4.8 (Remark) The bivector space $\mathcal{G}^2$, endowed with the commutator bracket, is a Lie algebra. It is obvious that the bracket is bilinear and skew-symmetric, and the Jacobi identity can be checked with the same reasoning as in E.5.10, p. 105.We are going to say that it is the Lie algebra of the rotor group $\mathcal{R}$, and we will write

$\mathfrak{lie}(\mathcal{R})$ to denote it. Now, we may ask whether the linear isomorphism $\mathsf{ad} : \mathcal{G}^2 \simeq \mathfrak{so}(E)$ is in fact an isomorphism of Lie algebras. This is indeed the case, since an easy calculation shows that $\mathsf{ad}_{[b,b']} = [\mathsf{ad}_b, \mathsf{ad}_{b'}]$, and this can be summed up as an isomorphism:

$$\mathcal{G}^2 = \mathfrak{lie}(\mathcal{R}) \simeq \mathfrak{lie}(\mathrm{SO}(E)) = \mathfrak{so}(E).$$

5.4.9 (Remark: The Differential of ρ at 1) In terms of multivariable calculus, the definition of an infinitesimal rotation coincides with the notion of tangent vector to $\mathrm{SO}(E)$ at the Id (▷ 6.5.1, 6.5.3, 6.5.5). In notation that is self-explanatory, we may write

$$T_{\mathrm{Id}}\mathrm{SO}(E) = \mathfrak{so}(E).$$

Similarly, the definition of infinitesimal rotor agrees with the notion of tangent vector to $\mathcal{R}$ at 1, and thus we may write

$$T_1(\mathcal{R}) = \mathcal{G}^2.$$

Now, we may consider the differential of the representation $\rho : \mathcal{R} \to \mathrm{SO}^0$ at 1 (we do not need $\tilde{\rho}$ because $\mathcal{R} \subseteq \mathcal{G}^+$). It is a linear map:

$$d_1\rho : T_1\mathcal{R} \to T_{\mathrm{Id}}\mathrm{SO}.$$

By definition, to get the image of a tangent vector b to $\mathcal{R}$ at 1 (an infinitesimal rotor), we have to take a path $R(t)$ on $\mathcal{R}$ such that $R'(0) = b$ and find the infinitesimal rotation associated to the path $\rho(R(t))$. Since we can take $R(t) = e^{tb}$, and

$$\rho\big(e^{tb}\big)(\boldsymbol{x}) = e^{tb}\boldsymbol{x}e^{-tb} = e^{t\,\mathsf{ad}_b}(\boldsymbol{x}),$$

we get that $(d_1\rho(b))(\boldsymbol{x}) = \mathsf{ad}_b(\boldsymbol{x})$, which shows that

$$d_1\rho = \mathsf{ad}.$$

In particular, $d_1\rho$ is an isomorphism.

5.4.10 (The Differential of the Exponential) Consider the *exponential map* of $\mathcal{R}$, namely $\exp : \mathcal{G}^2 \to \mathcal{R}$, $b \mapsto e^b$. Since $\mathcal{G}^2$ is linear, we can identify $T_0\mathcal{G}^2 = \mathcal{G}^2$. So, we get a linear map:

$$d_0 \exp : \mathcal{G}^2 = T_0\mathcal{G}^2 \to T_1\mathcal{R} = \mathcal{G}^2.$$

This map is the identity, as we can take $t \mapsto tb$ as a path on $\mathcal{G}^2$ with tangent vector b and then

$$(d_0 \exp)(b) = \tfrac{d}{dt}e^{tb}\big|_{t=0} = b.$$

Now, the inverse function theorem tells us that there exists an open neighborhood of 0 in $\mathcal{G}^2$ that exp maps diffeomorphically to an open neighborhood U of 1 in $\mathcal{R}$ (▷ 6.5.6). For rotors R in U, there exists a bivector b such that $u = e^b$, but for arbitrary rotors we can only assure that they will be a product of exponentials of bivectors (we will prove this in next Remarks).

Similar arguments work for $\mathrm{SO}(E)$: The exponential map:

$$\exp : T_0\mathfrak{so}(E) = \mathfrak{so}(E) \to T_{\mathrm{Id}}\mathrm{SO}(E) = \mathfrak{so}(E)$$

is the identity, and so there is an open neighborhood of 0 in $\mathfrak{so}(E)$ that exp maps diffeomorphically to an open neighborhood V of $\mathrm{Id} \in \mathrm{SO}(E)$. For rotations f in V, there exists a skew-symmetric endomorphism h such that $f = e^h$, but for arbitrary rotations in SO^0 we only can say that they will be a product of exponentials of skew-symmetric endomorphisms.

5.4.11 (Remarks) Assume that $\mathcal{R}$ is connected, and hence SO^0 is also connected. Then, any rotor has the form $e^{b_1} \cdots e^{b_k}$, for some $k \geqslant 1$ ($b_1, \ldots, b_k \in \mathcal{G}^2$), and any rotation has the form $e^{h_1} \cdots e^{h_l}$, for some $l \geqslant 1$ ($h_1, \ldots, h_l \in \mathfrak{so}(E)$). In both cases, we know that there exists an open neighborhood U of the identity (1 in the case of $\mathcal{R}$ and Id in the case of SO) whose elements are exponentials (e^b, $b \in \mathcal{G}^2$, in the case of $\mathcal{R}$, and e^h, $h \in \mathfrak{so}$, in the case of SO). Since the proof is similar, and valid in other similar contexts, let us concentrate on one, say $\mathcal{R}$. We may assume that the open set U is connected and symmetric, meaning that $R \in U \Leftrightarrow R^{-1} = \tilde{R} \in U$. In this way, the group generated by U is the union $\cup_{m \geqslant 1} U^m$, where U^m is the image of the multiplication map $U \times \cdots \times U \to \mathcal{R}$. Since $U^m = \cup_{R \in U} RU^{m-1}$, it is immediate to conclude by induction that U^m is open for all m (we use that the map $\mathcal{R} \to \mathcal{R}$, $X \mapsto RX$, is continuous, one-to-one, and with inverse which is also continuous). This means that the group, say G, generated by U is open, since the union of open sets is open. Now, $\mathcal{R}$ is the union of G and all the cosets RG for $R \notin G$. Since each of these cosets is open, it turns out that $\mathcal{A}\backslash G$ is open and hence G is also closed. So, we must have $G = \mathcal{R}$ because $\mathcal{R}$ is connected.

The problem of whether any rotor (any rotation) is the exponential of a single bivector (skew-symmetric endomorphism) is rather involved, and in general the answer is negative. Here, we are going to consider only the case of Euclidean spaces, which has a positive answer, and refer to other results in the literature (▷ 6.5.10) for readers wishing to know more about this question. Our main tool for the Euclidean case is the structure of normal endomorphisms, to which we turn in next subsection. We get at the same time the structure of two of the classes we will need (isometries and skew-endomorphisms), and also a proof that symmetric endomorphisms are diagonizable (in an orthonormal basis).

Normal Endomorphisms of E_n

Let $E = E_n$ and fix a normal $f \in \mathrm{End}(E)$. For any univariate polynomial P, we know that $P(f^\dagger) = P(f)^\dagger$, and we will simply write P and $P^\dagger$ to denote them. In case we need to refer to P as a polynomial, we will make it clear in some way, like by writing $P(X)$, where X is the polynomial variable.

5.4.12 (Remark) For any polynomial P, and any vectors $\boldsymbol{x}, \boldsymbol{y} \in E$,

$$(P\boldsymbol{x}) \cdot (P\boldsymbol{y}) = \boldsymbol{x} \cdot (P^\dagger P\boldsymbol{y}) = \boldsymbol{x} \cdot (PP^\dagger \boldsymbol{y}) = (P^\dagger \boldsymbol{x}) \cdot (P^\dagger \boldsymbol{y}).$$

In particular, $(f\boldsymbol{x})\cdot(f\boldsymbol{y}) = (f^\dagger \boldsymbol{x})\cdot(f^\dagger \boldsymbol{y})$. Setting $\boldsymbol{y} = \boldsymbol{x}$, we immediately conclude that $\ker P = \ker P^\dagger$.

Let Q be the minimal polynomial of f. Recall that Q is the unique monic polynomial such that $Q\boldsymbol{x} = 0$ for all $\boldsymbol{x} \in E$, and which divides any other polynomial having this property (*monic* means that the coefficient of highest degree is 1).

5.4.13 (Form of Q)

(1) *If P is an irreducible divisor of Q, then the multiplicity of P in Q is* 1 (in other words, P^2 does not divide Q).
(2) *The polynomial Q has the form*:

$$(X - \lambda_1)\cdots(X - \lambda_k)\big(X^2 - 2b_1 X + a_1^2\big)\cdots\big(X^2 - 2b_l X + a_l^2\big),$$

where $\lambda_i, a_j, b_j \in \mathbb{R}$ and $a_j > |b_j|$ $(i = 1, \ldots, k,\ j = 1, \ldots, l,\ k, l \geqslant 0,$ $k + 2l \leqslant n)$.

Proof

(1) If P^2 divides Q, there is a non-zero vector $\boldsymbol{x}$ such that $P^2\boldsymbol{x} = 0$ but $P\boldsymbol{x} \neq 0$. But, this leads to a contradiction, because on one hand we have

$$(P^\dagger P\boldsymbol{x}) \cdot (P^\dagger P\boldsymbol{x}) = (P\boldsymbol{x}) \cdot (PP^\dagger P\boldsymbol{x}) = (P\boldsymbol{x}) \cdot (P^\dagger P^2\boldsymbol{x}) = 0,$$

so $P^\dagger P\boldsymbol{x} = 0$, and on the other $(P\boldsymbol{x}) \cdot (P\boldsymbol{x}) = \boldsymbol{x} \cdot (P^\dagger P\boldsymbol{x}) = 0$, which implies $P\boldsymbol{x} = 0$.
(2) Let P be an monic irreducible factor of Q, and λ a complex root of P (it exists by the fundamental theorem of algebra). If $\lambda \in \mathbb{R}$, then $P = X - \lambda$. If $\lambda \notin \mathbb{R}$, then the complex conjugate $\bar{\lambda}$ is also a root, because P has real coefficients, and hence P is divisible by $(X - \lambda)(X - \bar{\lambda}) = X^2 - 2bX + a^2$, where $2b = \lambda + \bar{\lambda}$ and $a^2 = \lambda\bar{\lambda}$ $(a > 0)$. Since $X^2 - 2bX + a^2$ has real coefficients, we in fact have $P = X^2 - 2bX + a^2$. The condition $a > |b|$ is equivalent to say that P has no real roots, and $k + 2l \leqslant n$ is necessary because the degree of Q does not exceed n, as it divides the characteristic polynomial of f. □

5.4.14 (Main Reduction) *Let $P_1, \ldots, P_m$ be the monic irreducible factors of Q and set $F_j = \ker P_j$. Then, $E = F_1 + \cdots + F_m$ and $F_i \perp F_j$ when $i \neq j$.*

Proof If $m = 1$, there is nothing to prove. So, assume $m \geqslant 2$ and set $P = P_1$ and $Q' = P_2 \cdots P_m$. Since P and Q' are relatively prime, we can find polynomials W, W' such that $WP + W'Q' = 1$ (see E.5.13, p. 106). Then for any vector $\boldsymbol{x}$, we have $\boldsymbol{x} = \boldsymbol{x}_1 + \boldsymbol{x}'$, where $\boldsymbol{x}_1 = W'Q'\boldsymbol{x}$ and $\boldsymbol{x}' = WP$. Since $\boldsymbol{x}_1 \in \ker P = F_1$ and $\boldsymbol{x}' \in E' = \ker(Q')$, we have $E = F_1 + E'$. We also have $F_1 \cap E' = \{0\}$ (because for a vector $\boldsymbol{x}$ in this intersection we obviously have $\boldsymbol{x}_1 = \boldsymbol{x}' = 0$) and $F_1 \perp E'$. For the latter relation, first note that P induces an automorphism of E', as E' is invariant and we just have seen that non-zero vectors of E' cannot be in the kernel of P. So, there is a unique $\boldsymbol{y}'$ such that $\boldsymbol{x}' = P\boldsymbol{y}'$ and $\boldsymbol{x}_1 \cdot \boldsymbol{x}' = \boldsymbol{x}_1 \cdot (P\boldsymbol{y}') = (P^\dagger \boldsymbol{x}_1) \cdot \boldsymbol{y}' = 0$, where in the last step we use 5.4.12. Now, the proof follows by induction, because Q' is the minimal polynomial of E'. □

The main reduction leaves us with the analysis of the case in which Q is irreducible. If Q is linear, that $f = \lambda\mathrm{Id}$. Note that in the skew-symmetric case we must have $\lambda = 0$ and in the orthogonal case $\lambda = \pm 1$. So, we are left with the case in which $Q = X^2 - 2bX + a^2$, with $a, b \in \mathbb{R}$ and $a > |b|$. Note that in the skew-symmetric case we must have $b = 0$, as $Q(-X) = X^2 + 2bX + a^2$, the minimum polynomial of $-f$, has to agree with Q, and that in the orthogonal case we must have $a = 1$, as the monic polynomial $a^{-2}X^2(\frac{1}{X^2} - 2b\frac{1}{X} + a^2)$, which is the minimum polynomial of $f^\dagger = f^{-1}$, has to coincide with Q (hence $a^{-2} = a^2$).

Pick any unit vector $\boldsymbol{u}$ and a unit vector $\boldsymbol{v}$ in the plane $\langle \boldsymbol{u}, f\boldsymbol{u}\rangle$ (note that this is possible because f does not have real eigenvalues). Then, $f\boldsymbol{u} = \alpha\boldsymbol{u} + \beta\boldsymbol{v}$ ($\alpha, \beta \in \mathbb{R}$), and without loss of generality we may assume that $\beta > 0$ (replacing $\boldsymbol{v}$ by $-\boldsymbol{v}$ changes β to $-\beta$). Then, we have

5.4.15 *The plane $U = \langle \boldsymbol{u}, \boldsymbol{v}\rangle$ is invariant by f, $\alpha = b$, $\beta = \sqrt{a^2 - b^2}$, and the matrix of the restriction $f|_U$ of f to U, with respect to the basis $\{\boldsymbol{u}, \boldsymbol{v}\}$, is* $\begin{pmatrix} b & -\beta \\ \beta & b \end{pmatrix}$.

Proof We can easily find two expressions of $f^2\boldsymbol{u}$ using the relations $f\boldsymbol{u} = \alpha\boldsymbol{u} + \beta\boldsymbol{v}$ and $f^2 = 2bf - a^2\mathrm{Id}$ (that is, $Q = 0$):

$$f^2\boldsymbol{u} = \begin{cases} f(\alpha\boldsymbol{u} + \beta\boldsymbol{v}) = \alpha(\alpha\boldsymbol{u} + \beta\boldsymbol{v}) + \beta(f\boldsymbol{v}) = \alpha^2\boldsymbol{u} + \alpha\beta\boldsymbol{v} + \beta(f\boldsymbol{v}) \\ 2bf(\boldsymbol{u}) - a^2\boldsymbol{u} = 2b\big(\alpha\boldsymbol{u} + \beta\boldsymbol{v}\big) - a^2\boldsymbol{u} = (2b\alpha - a^2)\boldsymbol{u} + 2b\beta\boldsymbol{v}. \end{cases}$$

From this, we obtain

$$f\boldsymbol{v} = \gamma\boldsymbol{u} + (2b - \alpha)\boldsymbol{v}, \quad \gamma = \big(2\alpha b - a^2 - \alpha^2\big)/\beta.$$

This shows that U is invariant by f, and that the matrix of $f|_U$ with respect to $\{\boldsymbol{u}, \boldsymbol{v}\}$ is $A = \begin{pmatrix} \alpha & \gamma \\ \beta & 2b - \alpha \end{pmatrix}$, and hence that the matrix of $f^\dagger|_U$ is A^T. Now, we have

$$\alpha^2 + \beta^2 = (f\boldsymbol{u}) \cdot (f\boldsymbol{u}) = \big(f^\dagger\boldsymbol{u}\big) \cdot \big(f^\dagger\boldsymbol{u}\big) = \alpha^2 + \gamma^2$$

and therefore $\gamma = \varepsilon\beta$, $\varepsilon = \pm 1$. Since $X^2 - 2bX + 2b\alpha - \alpha^2 - \varepsilon\beta^2$ is the characteristic polynomial of A, it must agree with Q. Therefore,

$$b^2 - (2b\alpha - \alpha^2 - \varepsilon\beta^2) = (b - \alpha)^2 + \varepsilon\beta^2$$

must be negative (otherwise Q would have two real roots), which is only possible if $\varepsilon = -1$, that is $\gamma = -\beta$. Finally, a similar computation with the relation:

$$(f\boldsymbol{u}) \cdot (f\boldsymbol{v}) = \left(f^{\dagger}\boldsymbol{u}\right) \cdot \left(f^{\dagger}\boldsymbol{v}\right)$$

yields $\alpha = b$, and $a^2 = \det(A) = b^2 - \beta^2$ yields the formula for β. □

The final touch in determining the structure of f is to use the decomposition $E = U + U^{\perp}$. Indeed, since $U^{\perp}$ is invariant by f, by induction we conclude that $E = U_1 + \cdots + U_p$, where the $U_j = \langle \boldsymbol{u}_j, \boldsymbol{v}_j \rangle$ are invariant planes, with $U_i \perp U_j$ for $i \neq j$ $(j = 1, \ldots, p)$, and such that the matrix of f with respect to $\{\boldsymbol{u}_j, \boldsymbol{v}_j\}$ has the same form $\begin{pmatrix} b & -\beta \\ \beta & b \end{pmatrix}$. In addition, as noted before, we have $b = 0$ in the skew-symmetric case, and $a^2 = 1$ in the orthogonal case. In the latter, owing to $|b| < a^2 = 1$, there exists a unique angle $\theta \in (0, \pi)$ such that $b = \cos\theta$, and then $\beta = \sqrt{1 - b^2} = \sin\theta$, so that the previous matrix becomes $\begin{pmatrix} \cos\theta & -\sin\theta \\ \sin\theta & \cos\theta \end{pmatrix}$, the matrix of a rotation by θ in each of the planes U_j.

The next statement sums up the results, with indications about the form they take for symmetric, skew-symmetric, and orthogonal endomorphisms.

5.4.16 (Classification of Normal Endomorphisms of E_n)

(1) *If f is a normal endomorphism of the Euclidean space E_n, then its minimal polynomial has the form* 5.4.13.

(2) *If we set $D_i = \ker(X - \lambda_i)$ $(i = 1, \ldots, k)$ and $E_j = \ker(X^2 - 2b_jX + a_j^2)$ $(j = 1, \ldots, l)$, then $E = D_1 + \cdots + D_k + E_1 + \cdots + E_l$ is a decomposition as a sum of invariant subspaces, and two distinct summands are orthogonal.*

(3) *In D_i, the map f is $\lambda_i \mathrm{Id}_{D_i}$, and in E_j there is an orthonormal basis with respect to which the matrix of f has the form:*

$$\begin{pmatrix} b_j & -\beta_j & & & \\ \beta_j & b_j & & & \\ & & \ddots & & \\ & & & b_j & -\beta_j \\ & & & \beta_j & b_j \end{pmatrix},$$

where $\beta_j = \sqrt{a_j^2 - b_j^2}$.

(4) *For symmetric endomorphisms, the minimum polynomial of f only has linear factors, and therefore it takes diagonal form in an orthonormal basis.*

(5) *For skew-symmetric endomorphisms, the only possible real eigenvalue is* 0 *and $b_j = 0$ for all j.*

(6) *For orthogonal endomorphisms, the possible real values are* 1 *and* -1, *and there are angles $\theta_j \in (0, \pi)$ such that $b_j = \cos\theta_j$ and $\beta_j = \sin\theta_j$.*

5.4.17 (Remark on a Theorem of Cayley) From the classification above, we deduce that $1 \pm h$ is invertible for any skew-symmetric endomorphism. In fact, with the notations as in the last statement, and with $I = \mathrm{Id}$, we have

$$\det(I + h) = \beta_1^2 \cdots \beta_l^2.$$

Since $(I \pm h)^\dagger = I \mp h$, we conclude that

$$f = \frac{I - h}{I + h} \in \mathrm{SO}_n \quad \text{for all} \quad h \in \mathfrak{so}_n.$$

The isometry f does not have the eigenvalue -1: A vector $\boldsymbol{x}$ such that $f\boldsymbol{x} = -\boldsymbol{x}$ must satisfy $\boldsymbol{x} - h\boldsymbol{x} = -\boldsymbol{x} - h\boldsymbol{x}$ and so $\boldsymbol{x} = 0$.

From the last formula, we can solve for h in terms of f:

$$h = \frac{I - f}{I + f}.$$

It is to be noted that for any isometry f not having -1 as an eigenvalue we have (again with the notation as above)

$$\det(I + f) = \prod_j \left((1 + \cos\theta_j)^2 + \sin^2\theta_j\right) = \prod_j 2(1 + \cos\theta_j),$$

which is non-zero because $\theta_j \in (0, \pi)$, and that $h = (I - f)/(I + f)$ is skew-symmetric, because its adjoint is $(I - f^{-1})/(I + f^{-1}) = (f - I)/(f + I) = -h$.

The Exponential Maps for SO_n and $\mathcal{R}_n$

5.4.18 *The exponential map* $\exp : \mathfrak{so}_n \to \mathrm{SO}_n$ *is onto for all* $n \geqslant 1$. *In other words, for any* $f \in \mathrm{SO}_n$ *there exists* $h \in \mathfrak{so}_n$ *such that* $f = e^h$.

Proof This follows from 5.4.16 and the observation that the matrix:

$$A_\theta = \begin{pmatrix} \cos\theta & -\sin\theta \\ \sin\theta & \cos\theta \end{pmatrix}$$

is the exponential of

$$H_\theta = \begin{pmatrix} 0 & -\theta \\ \theta & 0 \end{pmatrix}$$

(see E.5.11, p. 105). Since this works for any value of θ, in particular we get, for $\theta = \pi$, that $-I_2$ is the exponential of H_π. With this, the proof is complete in view that the multiplicity of the eigenvalue -1 of an element of SO_n is necessarily even, and that 1 is the exponential of 0. □

5.4.19

(1) *For any* $f \in \mathrm{SO}_n$, *there exists* $b \in \mathcal{G}^2$ *such that*

$$f(\boldsymbol{x}) = e^b \boldsymbol{x} e^{-b} = e^{\mathrm{ad}_b}(\boldsymbol{x}) \text{ for all } \boldsymbol{x} \in E_n.$$

(2) *For any rotor* $R \in \mathcal{R}_n$, *either* R *or* $-R$ *has the form* e^b, $b \in \mathcal{G}^2$.

Proof

(1) By 5.4.18, there exists $b \in \mathcal{G}^2$ such that $\underline{R} = e^{\mathrm{ad}_b}$.
(2) Let $b \in \mathcal{G}^2$ be such that $\underline{R} = e^{\mathrm{ad}_b}$. Then, $R' = e^b$ satisfies that $\underline{R'} = e^{\mathrm{ad}_b} = \underline{R}$, and hence we must have $R \in \{\pm R'\} = \{\pm e^b\}$. □

5.5 Exercises

E.5.1 (The Algebra $\mathbb{C} \otimes \mathbb{H}$) Show that $\mathbb{C} \otimes \mathbb{H} \simeq \mathbb{C}(2)$.

Hint Given $z \in \mathbb{C}$ and $h \in \mathbb{H}$, define $f_{z,h} : \mathbb{H} \to \mathbb{H}$ by the formula $f_{z,h}(x) = zx\bar{h}$. Then, $f_{z,h}$ is $\mathbb{C}$-linear, or $f_{z,h} \in \mathrm{End}_{\mathbb{C}}(\mathbb{H})$. Since $(z, h) \mapsto f_{z,h}$ is bilinear, there is a unique linear map:

$$\mathbb{C} \otimes \mathbb{H} \to \mathrm{End}_{\mathbb{C}}(\mathbb{H})$$

such that $z \otimes h \mapsto f_{z,h}$. This map is in fact an algebra homomorphism, as follows from the computation $(f_{z',h'} \circ f_{z,h})(x) = z'(zx\bar{h})\bar{h}' = z'zx\overline{h'h} = f_{z'z,h'h}(x)$. Now, it is straightforward to check that the map sends the basis $\{1, i\} \otimes \{1, \boldsymbol{i}, \boldsymbol{j}, \boldsymbol{k}\}$ into linearly independent endomorphisms, and hence the map is an isomorphism, for both sides have dimension 8. Finally, note that $\mathrm{End}_{\mathbb{C}}(\mathbb{H}) \simeq \mathrm{End}_{\mathbb{C}}(\mathbb{C}^2) \simeq \mathbb{C}(2)$.

E.5.2 (The Algebra $\mathbb{H} \otimes \mathbb{H}$) Prove that $\mathbb{H} \otimes \mathbb{H} \simeq \mathbb{R}(4)$.

Hint Given $h_1, h_2 \in \mathbb{H}$, define $f_{h_1,h_2} : \mathbb{H} \to \mathbb{H}$ by the formula $f_{h_1,h_2}(x) = h_1 x \bar{h}_2$. In this way we get, as in E.5.1, an algebra homomorphism $\mathbb{H} \otimes \mathbb{H} \to \mathrm{End}(\mathbb{H})$ which can be shown to be an isomorphism (both sides have dimension 16). Finally, $\mathrm{End}(\mathbb{H}) \simeq \mathrm{End}(\mathbb{R}^4) \simeq \mathbb{R}(4)$.

E.5.3 (Alternative Definition of Skew-Symmetric Endomorphisms) Show that an endomorphism h of (E, q) is skew-symmetric (see Eq. (5.1) for the definition) if and only if $\boldsymbol{x} \cdot (h\boldsymbol{x}) = 0$ for all $\boldsymbol{x} \in E$. This is an unpolarized form of Eq. (5.1), in the sense that we get the present definition by setting $\boldsymbol{y} = \boldsymbol{x}$ in the former one.

E.5.4 Construct an $R \in \mathcal{G}_6^+$ such that $R\tilde{R} = 1$, but which is not a rotor.

Hint Try $R = \rho(1 + I)$, where I is a pseudoscalar and $\rho = 1/\sqrt{2}$.

E.5.5 (1) Prove that path-connected space is connected. (2) Conversely, prove that a connected and locally path-connected space is path-connected.

Hint (1) If X is path-connected, and $Y \subset X$ is a non-empty open and closed subset, for any path $\gamma : I = [0, 1] \to X$ such that $\gamma(0) \in Y$, $\gamma^{-1}(Y)$ is a non-empty open and closed subset of I and hence $\gamma(I) \subset Y$, as I is connected. Since we can choose γ so that $\gamma(1)$ is any chosen point of X, we conclude that $Y = X$ and so X is connected. (2) Fix a point $a \in X$, let X' be the subset of X formed by the points x that are path-connected to a, and show that X' and $Y \backslash X'$ are open, so $X' = X$ because X' is open and closed in X.

E.5.6 If $\boldsymbol{u}$ and $\boldsymbol{v}$ are orthogonal non-zero vectors of E_3, show that $s_{\boldsymbol{u}} s_{\boldsymbol{v}} = s_{\boldsymbol{u}\times\boldsymbol{v}} = s_{\boldsymbol{v}\times\boldsymbol{u}}$.

E.5.7 With the notations as for Eq. (5.5), show that the vectors $\boldsymbol{w}^{\pm} = \epsilon e^{\pm\alpha}\boldsymbol{u} - \boldsymbol{v}$ are isotropic.

Hint Compute $(\boldsymbol{w}^{\pm})^2$ and use that $\boldsymbol{u} \cdot \boldsymbol{v} = \cosh\alpha$. It is also easy to see that the expressions for $\boldsymbol{w}^{\pm}$ can be obtained by solving for λ the quadratic equation:

$$(\lambda\boldsymbol{u} - \boldsymbol{v})^2 = 0.$$

E.5.8 (Infinitesimal Rotations of E_2) Let $\boldsymbol{i}$ be the unit area of the Euclidean plane E_2 and ω a scalar. Then, the expression $f_{\omega,t}\boldsymbol{x} = \boldsymbol{x}e^{\boldsymbol{i}\omega t}$ yields the counterclockwise rotation of $\boldsymbol{x}$ by the angle ωt (circular motion of $\boldsymbol{x}$ of *angular velocity* ω). Thus, $f_{\omega,t} \in \mathrm{SO}_2$ is a differentiable path on SO_2 and since $f_{\omega,0} = \mathrm{Id}$, we can consider the associated infinitesimal rotation r_ω. In this case, we have

$$r_\omega(\boldsymbol{x}) = \omega\boldsymbol{x}\boldsymbol{i} = \omega\boldsymbol{x}^{\perp},$$

where $\boldsymbol{x}^{\perp}$ is the counterclockwise rotation of $\boldsymbol{x}$ by $\pi/2$. Indeed,

$$r_\omega(\boldsymbol{x}) = \tfrac{d}{dt}(\boldsymbol{x}e^{\boldsymbol{i}\omega t})\big|_{t=0} = \boldsymbol{x}\boldsymbol{i}\omega.$$

The fact that r_ω is skew-symmetric is also readily checked: since $\boldsymbol{x}\boldsymbol{i}$ is orthogonal to $\boldsymbol{x}$, $(r_\omega\boldsymbol{x}) \cdot \boldsymbol{x} = \omega(\boldsymbol{x}\boldsymbol{i}) \cdot \boldsymbol{x} = 0$. Note that in this case the space of skew-symmetric endomorphisms is isomorphic to $\{\omega\boldsymbol{i} \mid \omega \in \mathbb{R}\} = \mathbb{R}\boldsymbol{i}$, which is the $\boldsymbol{i}$-axis in the geometric complex plane $\mathbf{C} = \mathcal{G}_2^+$. Since it is one-dimensional, the commutator bracket is identically 0.

E.5.9

(1) The existence of $f^{\dagger}$ is an immediate consequence of the regularity of q. Note that fixing an arbitrary $\boldsymbol{y}$, the expression $f(\boldsymbol{x}) \cdot \boldsymbol{y}$ is linear in $\boldsymbol{x}$, and this linear form is achieved by the dot product with a unique vector $\boldsymbol{y}'$, that is, $(f\boldsymbol{x}) \cdot \boldsymbol{y} = \boldsymbol{x} \cdot \boldsymbol{y}'$ for all $\boldsymbol{x}$. Since $\boldsymbol{y}$ is arbitrary, we may consider the map $E \to E$, $\boldsymbol{y} \mapsto \boldsymbol{y}'$. Using the bilinearity of the dot product, it is easy to see that this map, which is denoted by $f^{\dagger}$, is linear (check it), and by construction it satisfies the Eq. (5.7).

Finally, we have

$$(gf\boldsymbol{x})\cdot\boldsymbol{y}=(f\boldsymbol{x})\cdot\left(g^{\dagger}\boldsymbol{y}\right)=\boldsymbol{x}\cdot\left(f^{\dagger}g^{\dagger}\boldsymbol{y}\right),\ \text{ but also }\ (gf\boldsymbol{x})\cdot\boldsymbol{y}=\boldsymbol{x}\cdot\left((gf)^{\dagger}\boldsymbol{y}\right),$$

and hence we indeed have $(gf)^{\dagger}=f^{\dagger}g^{\dagger}$.

(2) Let f be an endomorphism of E and A its matrix with respect to an orthonormal basis $\mathbf{e}=\boldsymbol{e}_1,\ldots,\boldsymbol{e}_n$. Show that the matrix of $f^{\dagger}$ with respect to the conjugate basis $\mathbf{e}$ is the matrix $B=A^{\dagger}$ defined as follows: $b_{ij}=\epsilon_{ij}a_{ji}$, where $\epsilon_{ij}=1$ if $1\leqslant i,j\leqslant r$ or $r+1\leqslant i,j\leqslant= r+s=n$, and $\epsilon_{ij}=-1$ otherwise. In a more visual way, if X (Y) is the upper-left (bottom-right) $r\times r$ ($s\times s$) submatrix of A, and P (Q) is the residual $r\times s$ ($s\times r$) upper-right (bottom-left) corner submatrix, then

$$A^{\dagger}=\begin{pmatrix}X & P\\ Q & Y\end{pmatrix}^{\dagger}=\begin{pmatrix}X^T & -Q^T\\ -P^T & Y^T\end{pmatrix}.$$

In particular, we have $A^{\dagger}=A^T$ when $r=n$ (Euclidean case) or $s=n$ (anti-Euclidean case).

(3) Show that the space of skew-symmetric endomorphisms has dimension $\binom{n}{2}$.

Hint The condition $A^{\dagger}=-A$ is equivalent to $X^T=-X$, $Y^T=-Y$, and $P^T=Q$, which leaves $\binom{r}{2}+\binom{s}{2}+rs$ free parameters.

E.5.10 (Every Associative Algebra Is a Lie Algebra) If $\mathcal{A}$ is any associative algebra, the commutator bracket $[x,y]$ of two elements $x,y\in\mathcal{A}$ is defined to be $xy-yx$. This bracket is bilinear (because the product of $\mathcal{A}$ is bilinear) and skew-symmetric (because $[x,x]=0$ for all $x\in\mathcal{A}$). Moreover, we have

$$[x,[y,z]]=x(yz-zy)-(yz-zy)x=xyz-xzy-yzx+zyx.$$

Permuting this identity cyclically and adding the three identities, we obtain the relation:

$$[x,[y,z]]+[y,[z,x]]+[z,[x,y]]=0.$$

In other words, the bracket satisfies the Jacobi identity, and hence $(\mathcal{A},[-,-])$ is a Lie algebra.

E.5.11 Given the matrix $H_\theta=\begin{pmatrix}0 & -\theta\\ \theta & 0\end{pmatrix}$, show that

$$H_\theta^{2k}=(-1)^k\theta^{2k}I_2\ \text{ and }\ H_\theta^{2k+1}=(-1)^k\theta^{2k}H_\theta.$$

Conclude that $e^{H_\theta}=\begin{pmatrix}\cos\theta & -\sin\theta\\ \sin\theta & \cos\theta\end{pmatrix}$ for all $\theta\in\mathbb{R}$.

E.5.12 (Analytic Functions) Let V be a finite dimensional vector space. A function $f : I \to V$, where $I = (\alpha, \beta) \subseteq \mathbb{R}$, is said to be (real) *analytic* if it has derivatives $D^k f = d^k f/dt^k$ for all $k > 0$ and the Taylor series $\sum_{k\geqslant 0} \frac{1}{k!} D^k f(t_0)(t-t_0)^k$ converges to f in an open neighborhood of t_0 for any $t_0 \in I$.

For example, the exponential function $e^t \in \mathbb{R}$ is defined for all $t \in \mathbb{R}$, and its Taylor series at any t_0 *converges to* e^t *for all* t. Indeed, we have $D^k e^t = e^t$ for all k, and hence the Taylor series of e^t at t_0 is $\sum_{k\geqslant 0} \frac{1}{k!} e^{t_0}(t-t_0)^k = e^{t_0} e^{t-t_0} = e^t$. The same argument works for the function $e^{tb} \in \mathcal{G}$, hence also for the function $f(t) = e^{tb}\boldsymbol{x}e^{-tb} \in \mathcal{G}$ considered in 5.4.6, and this justifies the claim that $e^{tb}\boldsymbol{x}e^{-tb} = e^{t\,\mathrm{ad}_b}(\boldsymbol{x})$ used there.

E.5.13 (Bézout Identity) In the case of integers, if d is the greatest common divisor of $m, n \in \mathbb{Z}$, $d = \gcd(m, n)$ in symbols, there exist $a, b \in \mathbb{Z}$ such that $d = am + bn$. This can be easily proven as follows. We can assume that $m \geqslant n \geqslant 0$. If $n = 0$ or $n = m$, then $d = m$ and it is enough to take $a = 1$, $b = 0$. So, we may assume that $m > n > 0$. Since $\gcd(m-n, n) = \gcd(m, n) = d$, by induction on the greatest of the two numbers we can find $a', b' \in \mathbb{Z}$ such that $d = a'(m-n) + b'n = a'm + (b'-a')n$ and so it is enough to take $a = a'$ and $b = b' - a'$.

A quicker way to get a and b is to observe that if $m = qn + r$ is the Euclidean division of m by n, so that $0 \leqslant r < n$, then $\gcd(n, r) = \gcd(n, m - qn) = d$ and so, by induction, there exist $a', b' \in \mathbb{Z}$ such that

$$d = a'n + b'r = a'n + b'(m - qn) = b'm + (a' - b'q)n$$

and so it is enough to take $a = b'$ and $b = a' - b'q$.

The latter procedure tells us that there is a Bézout identity for two univariate polynomials, not both zero, because we can use the Euclidean division of univariate polynomials and argue by induction on the maximum degree of the two polynomials in question.

Chapter 6
Postfaces

6.0 Prelude

This chapter is a collection of notes. Most of them are pointed out at suitable spots in the text of the preceding chapters, but an effort has been aimed to ensure that it may be read on its own as a closing chapter. The numbered sections, 6.1–6.5, correspond to Chaps. 1–5, respectively, while this prelude (6.0) corresponds to the Preface.

6.0.1 (*In Memoriam* of Waldyr A. Rodrigues Jr.) The story that led to the author's involvement with the topics studied in this book, and in the companion brief [98], was summarized in the lecture delivered at ICCA11 (Ghent, 07–11 August 2017) with the title *Waldyr Alves Rodrigues Jr.: Sketches on his Life and Work* (prepared jointly with Carlile Lavor).

The idea of proposing to speak about the life and work of W.A. Rodrigues Jr. took form early in the June 2017 visit of Carlile Lavor (University of Campinas, São Paulo, Brazil) to Barcelona. The visit had been planned almost 1 year before with two main goals: helping him in the organization of AGACSE 2018 in Campinas (July 23–27) and to do some work in the applications of Geometric Algebra to various questions. The main trail that led us to that point started, at least formally, with AGACSE 2015 (Barcelona, July 27–31),[1] of which the author chaired the Organizing Committee and was the Guest Editor in Chief of the Proceedings (see [150]). In that visit, we started to work on [98], and it also took form the idea of writing this brief.

6.0.2 (J.B. Sancho, an Early Teacher) In the ICCA11 lecture, the author outlined briefly a few involvements that preceded AGACSE 2015. "We studied Emil Artin's *Geometric Algebra* in 1966–1967 at the Faculty of Mathematics of the University of Barcelona, following lectures by Professor Juan Bautista Sancho Guimerá. He

[1] https://mat-web.upc.edu/people/sebastia.xambo/agacse2015/.

S. Xambó-Descamps, *Real Spinorial Groups*, SpringerBriefs in Mathematics,
https://doi.org/10.1007/978-3-030-00404-0_6

[Sancho] was as passionate about mathematics and physics as Waldyr, he also supervised many students (10 years in Barcelona, until 1972, and 40 in Salamanca, where he moved in that year, until his death in 2012), and both could concentrate for hours on their current topic, but otherwise they were completely opposite in one respect: whereas Waldyr was always extremely diligent to publish, Sancho's ideas were only, for the most part, communicated orally to his students."[2]

6.0.3 (The K2 Project) Two lectures were also referred. One was the 2009 Funchal lecture *A Clifford view of Klein's Geometry* [146] in one of the earliest international conferences on "The Klein Project,"[3] whose goal was set to produce, through a collective effort, an update for our times of Klein's *Geometrie* (the second volume of his *Elementarmathematik vom höheren Standpunk aus*). After having taught Geometry and Mathematical Models of Physics for many years at the Facultat de Matemàtiques i Estadística of the Universitat Politècnica de Catalunya, of having carried research on stereoscopic vision for a Barcelona company (Imagsa Technologies), of having discovered David Hestenes' teachings through many discussions with Josep M. Parra (University of Barcelona), and surely moved by the fact that W.K. Clifford had died in Funchal in 1879, I phrased my job as follows: "In the talk we will present, after a brief outline of the main features of the book [*Geometrie*, *Geometry* in the English translation] and of how it can account for 'newer' geometries, such as the special theory of relativity, a proposal of guidelines for its updating using geometric algebra (GA) will be made. First introduced by Clifford, as a generalization of Grassmann's algebra and Hamilton's quaternions, geometric algebra has developed into a unified basis for geometry and its applications, not only across pure mathematics, but also in physics, engineering and computation." About Hestenes contributions, it was said that "The honing of GA (in the full sense of Grassmann and Clifford) as a viable basis for a paradigm shift has been championed in the last 40 years by David Hestenes, a physicist/mathematician at Arizona State University, with important help and insights from others," and the main papers by him used for the occasion were:

- *Reforming the Mathematical Language of Physics* (Oersted Medal Lecture 2002, 43 pages)
- *Spacetime Physics with Geometric Algebra* (Amer. J. Physics, 2003, 24 pages)
- *Gauge theory gravity with Geometric Calculus* (Found. Phys., 2005, 67 pages)

In a remark at the end, it was observed that "Just as F. Klein does not mention Clifford, the descendants of Clifford seldom mention Klein, an exception being the Hestenes–Sobczyk book *Clifford Algebra to Geometric Calculus* [81, p. xi]."

6.0.4 (*In Memoriam* of J.B. Sancho) The other lecture mentioned at ICCA11 was the one delivered in April 2014 in the Symposium in memoriam of Sancho that lead

[2]Fortunately, part of Sancho's teachings have recently been systematically written up by Juan Antonio Navarro [111].

[3]https://www.mathunion.org/icmi/activities/klein-project.

to the paper [145]. The title of the talk, and of the paper published in the Proceedings volume, was (translated to English) *Hidden pathways from projective geometry to quantum physics*, with the third part devoted to GA. The "hidden pathways" phrase points to the poem "Vida retirada/Secluded life" of Fray Luis de León, a sixteenth century Augustinian friar, theologian and academic admired by Sancho. After imprisonment for 4 years in Valladolid by the Inquisition, Fray Luis began his first class in Salamanca on 29 January 1577 with the words *dicebamus hesterna die...* ("As we were saying yesterday..."). For the part on GA, I approached Manuel Parra (once more) and Ramon González to seek their advice. "This was just in the Spring before the Tartu ICCA and they managed – it was not that hard – to convince me to join them in organization of AGACSE 2015. I put all I could, institutionally and personally, for meeting that responsibility as best as possible."

6.0.5 (Other Lectures and Courses) Other lectures and short courses that have contributed to the writing of this brief are the following. (1) Intensive course (10 h) on *Geometric Algebra Techniques in Mathematics and Physics*, University of San Luis Potosí (Mexico), 9–13 March 2015. The course was extended with two lectures on spinors in two "Encuentros/Meetings" (18–23 March 2015), one in San Luis Potosí and another in San Carlos, Guaymas, Sonora, Mexico. (2) Two intensive courses on *Geometric Algebra Techniques in Mathematics, Physics and Engineering* (16–19 November 2015 and 16–20 May 2016) at the Mathematics Institute of the Universidad de Valladolid. The topics were similar to those dealt with in Mexico, but with more mathematical detail and including Conformal Geometric Algebra. (3) Intensive course (10 h) on *Formalisms in Mathematical Physics: Perspectives, Structures and Methods* at the University of San Luis Potosí (June 27 to July 1, 2016). (4) Four lectures at the 17th Lluís Santaló Research School (22–26 August 2016, Santander)[4] and the lecture *Dirac's equation seen in the GA light* at the Barcelona CSASC Joint Meeting (20–23 September 2016).[5]

6.1 Topology Background

In this section, we recall some general topological ideas and facts, including the notion of homotopy between continuous maps; the fundamental group encountered in 5.1.5, p. 80; and a description of the fundamental group of $\mathrm{SO}^0_{r,s}$. As a general and accessible reference, see [127], particularly Chap. 3. Let us also note that Topology is increasingly showing its relevance in many fronts of current Physics' research as surveyed in recent works such as [5, 26], and [78, 79].

6.1.1 (Basic Notions) For our purposes, we only need to consider topological spaces that are subsets of some finite-dimensional real vector space, which we will simply call *spaces* (with the exception of projective spaces and Grassmannians

[4] https://mat-web.upc.edu/people/sebastia.xambo/santalo2016/.

[5] https://mat-web.upc.edu/people/sebastia.xambo/GA/2016-Xambo-csasc.pdf.

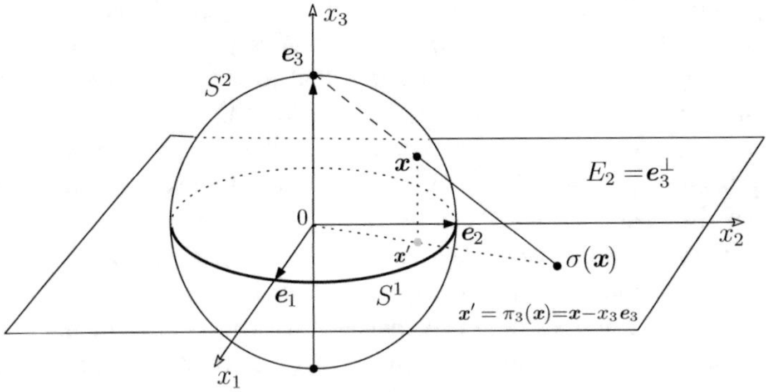

Fig. 6.1 Stereographic projection of $S^2\backslash\{\boldsymbol{e}_3\}$ to E_2 from $\boldsymbol{e}_3$. Analytically, $\sigma(\boldsymbol{x}) = \lambda\boldsymbol{x}'$, where $\boldsymbol{x}' = \boldsymbol{x} - x_3\boldsymbol{e}_3 = \pi_3\boldsymbol{x}$ (the orthogonal projection of $\boldsymbol{x}$ to E_2) and $\lambda = 1/(1 - x_3)$

discussed in 6.5.4). The topology of any such space is the topology induced by the standard topology of the vector space containing it.[6] Thus, an *open set* of $X \subseteq E$ is any subset U of X of the form $U = V \cap X$, where V is open in E. The *closed sets* of X are the complements of open sets.

A map $f : X \to X'$ between spaces is said to be *continuous* if $f^{-1}U'$ is an open set of X for any open set U' of X'. It is immediate to check that the composition of continuous maps is continuous. If f is bijective and f^{-1} is also continuous, we say that f is a *homeomorphism*. This is equivalent to say that $U \subset X$ is open in X if and only if $f(U)$ is open in X'.

6.1.2 (Stereographic Projection) Consider the sphere S^{n-1} of radius 1 in E_n:

$$S^{n-1} = \{\boldsymbol{x} \in E_n \mid \boldsymbol{x}^2 = 1\}.$$

Then, $\boldsymbol{e}_n \in S^{n-1}$ and the *stereographic projection* from $\boldsymbol{e}_n$ is the map:

$$\sigma : S^{n-1}\backslash\{\boldsymbol{e}_n\} \to E_{n-1}, \quad E_{n-1} = \boldsymbol{e}_n^\perp,$$

defined by requiring that $\sigma(\boldsymbol{x}) \in E_{n-1}$ be aligned with $\boldsymbol{e}_n$ and $\boldsymbol{x}$. See Fig. 6.1 for an illustration of the case $n = 3$. With a little analytic geometry, we find that

$$\sigma(\boldsymbol{x}) = \boldsymbol{x}'/(1 - x_n), \quad \boldsymbol{x}' = \boldsymbol{x} - x_n\boldsymbol{e}_n = \pi_n(\boldsymbol{x}), \tag{6.1}$$

[6] It is the topology induced by any *Euclidean norm* $||\boldsymbol{v}||$ on E, an (auxiliary) notation that we will use henceforth. One way of choosing an auxiliary norm is by transferring the standard norm of $\mathbb{R}^n$ by the isomorphism $E \simeq \mathbb{R}^n$ induced by a basis of E. In practical terms, this means that to take the limit of a sequence of vectors in E is reduced to taking the limits in $\mathbb{R}$ of the components of the vectors in the sequence. To avoid any confusion, recall that the *magnitude* $|\boldsymbol{v}|$ of $\boldsymbol{v}$ relative to a metric q (of arbitrary signature) is defined to be $\sqrt{\varepsilon_{\boldsymbol{v}} q(\boldsymbol{v})}$. Only if q is positive definite can we use the magnitude also as an auxiliary norm.

where π_n denotes the orthogonal projection onto E_{n-1}. In particular, we see that σ is continuous. But, in fact it is a homeomorphism, for $\boldsymbol{x}$ can be obtained from $\boldsymbol{y} = \sigma(\boldsymbol{x})$ by the formula:

$$\boldsymbol{x} = 2\boldsymbol{y}/(1+\boldsymbol{y}^2) + \boldsymbol{e}_n(\boldsymbol{y}^2-1)/(\boldsymbol{y}^2+1), \tag{6.2}$$

which is a continuous function of $\boldsymbol{y}$. Indeed, there is a non-zero scalar μ such that $\boldsymbol{x} = \boldsymbol{e}_n + \mu(\boldsymbol{y} - \boldsymbol{e}_n) = (1-\mu)\boldsymbol{e}_n + \mu\boldsymbol{y}$, and squaring we get, using $\boldsymbol{x}^2 = 1$ and $\boldsymbol{e}_n \cdot \boldsymbol{y} = 0$, $\mu = 2/(1+\boldsymbol{y}^2)$, from which the claimed relation follows.

6.1.3 (Homotopies) Two continuous maps $f, g : X \to X'$ are said to be *homotopic*, and we write $f \simeq g$ to denote it, if there is a continuous map $H : I \times X \to X'$, where $I = [0, 1] \subset \mathbb{R}$, such that

$$H(0, x) = f(x) \text{ and } H(1, x) = g(x) \text{ for all } x \in X.$$

To see that this expresses the idea of continuous deformation of f into g (or *homotopy*), consider the maps $h_s : X \to X'$, $s \in I$, defined by $h_s(x) = H(s, x)$. This is a continuously varying family $\{h_s\}_{s\in I}$ of continuous maps $h_s : X \to X'$ and by definition we have $h_0 = f$ and $h_1 = g$. The homotopy relation $\simeq$ turns out to be an equivalence relation in the set of maps $X \to X'$, and the *homotopy class* of f, consisting of all continuous maps $X \to X'$ that are homotopic to f, is denoted by $[f]$.

6.1.4 (Poincaré's Fundamental Group) Given a space X and a point $x_0 \in X$, the elements of the *fundamental group* of X with base point x_0, which is denoted by $\pi_1(X, x_0)$, are the homotopy classes $[\gamma]$ of *loops* on X with base point x_0, by which we mean continuous maps $\gamma : I \to X$ such that $\gamma(0) = \gamma(1) = x_0$. In this case, a homotopy $H : I \times I \to X$ is required to satisfy $H(s, 0) = x_0 = H(s, 1)$ for all $s \in I$, which means that all the paths $\gamma_s(t) = H(s, t)$ have to be loops on X at x_0 (*loop homotopy*). The group operation is defined by the rule $[\gamma][\gamma'] = [\gamma * \gamma']$, where $\gamma * \gamma'$ is the loop defined by:

$$(\gamma * \gamma')(t) = \begin{cases} \gamma(2t) & \text{for } 0 \leqslant t \leqslant \frac{1}{2}, \\ \gamma'(2t-1) & \text{for } \frac{1}{2} \leqslant t \leqslant 1. \end{cases}$$

Note that this loop travels the whole loop γ for $t \in [0, \frac{1}{2}]$ followed by traveling the whole loop γ' for $t \in [\frac{1}{2}, 1]$. The composition $\gamma * \gamma'$ is not associative, but it becomes so at the level of homotopy classes. Similarly, the constant loop $e : I \to X$, $e(t) = x_0$ for all t, is not a neutral element for the composition, but it is so for homotopy classes, namely $[e][\gamma] = [\gamma][e] = [\gamma]$; and the inverse loop γ^{-1} defined by traveling γ backwards, $\gamma^{-1}[t] = \gamma[1-t]$, satisfies $[\gamma][\gamma^{-1}] = [\gamma^{-1}][\gamma] = [e]$ although $\gamma * \gamma^{-1} \neq e$.

A continuous map $f : X \to X'$ induces a group homomorphism:

$$\tilde{f} : \pi_1(X, x_0) \to \pi_1(X', x_0'), \quad \text{where } x_0' = f(x_0).$$

Actually, if γ is a loop on X at x_0, then $\gamma' = f \circ \gamma$ is a loop on X' at x_0' and the homomorphism is defined by $[\gamma] \mapsto [\gamma']$. In particular, we see that if f is a homeomorphism, then $\tilde{f}$ is an isomorphism.

If $x_0, x_0' \in X$ are connected by a path δ, then the map $\pi_1(X, x_0') \to \pi_1(X, x_0)$, $[\gamma] \mapsto [\delta][\gamma][\delta^{-1}]$ is an isomorphism of groups, with inverse the analogous map for δ^{-1}. In particular, we see that for path-connected spaces the isomorphism class of $\pi_1(X, x_0)$ is the same for all points x_0. In such cases, we may simply write $\pi_1(X)$ to denote that isomorphism class. This is especially apt when X has some distinguished point, and of course also when $\pi_1(X) \simeq \{0\}$.

6.1.5 (Simply Connected Spaces) The space X is *simply connected* if and only if it is connected and π_1 is trivial. A vector space E is simply connected, as

$$H(s, t) = (1 - s)\gamma(t)$$

is a loop homotopy of any given loop γ on E at 0 to the constant loop at 0. The same argument works for *star-shaped* sets X, which by definition include the segment $\{t\boldsymbol{x}\}_{0 \leqslant t \leqslant 1}$ for all $\boldsymbol{x} \in X$.

The spheres S^{n-1} are simply connected for $n \geqslant 3$, as in this case any loop on S^{n-1} can be deformed to a loop that avoids $\boldsymbol{e}_n$ and hence

$$\pi_1(S^{n-1}) = \pi_1(S^{n-1} \backslash \{\boldsymbol{e}_n\}) = \pi_1(E_{n-1}) = \{0\}.$$

This argument does not work for S^1 ($n = 2$), for any loop on S^1 going at least once round it cannot be deformed to avoid $\boldsymbol{e}_2$. Actually, in this case $\pi_1(S^1) \simeq \mathbb{Z}$, where the isomorphism is given by counting the number of times a loop on S^1 goes round S^1, with the sign $\pm$ determined by the sense (counterclockwise or clockwise) of the net number of turns. This argument is made precise in next block.

6.1.6 (Topological Groups) As a general fact, all groups G that we have studied are *topological groups*, which means that the group operation $G \times G \to G$ and the inverse map $G \to G$, $g \mapsto g^{-1}$, are continuous.

For example, the group of $\mathrm{GL}(E)$ of linear automorphisms of a vector space E is a topological group. In fact, it is isomorphic to the group GL_n of (real) invertible matrices of order n, and in this group the expressions for the product of two matrices and for the inverse of a matrix clearly show that they are continuous maps. From this, it follows that any subgroup of $\mathrm{GL}(E)$ is a topological group with the induced topology. In particular, the orthogonal groups $\mathrm{O}_{r,s}$, $\mathrm{SO}_{r,s}$, and $\mathrm{SO}^0_{r,s}$ are topological groups.

A similar argument works for the spinorial groups. The geometric product of a geometric algebra $\mathcal{G}_{r,s}$ is continuous (as it is bilinear), and this implies that the product of $\mathcal{G}_{r,s}^{\times}$ is a continuous map. With a bit more work, we could show directly that the inverse map is also continuous, and hence $\mathcal{G}_{r,s}^{\times}$ is a topological group, but instead we appeal to the (independent) proof given in 6.5.8 using differential methods. Finally, the versor, pinor, spinor, and rotor groups are also topological groups with the topology induced by topology of $\mathcal{G}_{r,s}^{\times}$.

6.1.7 (The Fundamental Group of $\mathrm{SO}_{r,s}^0$ and $\mathcal{R}_{r,s}$) A subset Z of a space X is said to be *discrete* if for any $z \in Z$ there is an open set U in X such that $U \cap Z = \{z\}$. Any finite subset, for example, is discrete.

Let G be a connected group and Z a discrete subgroup. Let G/Z be the set of right Z-cosets and $p : G \to G/Z$ the map $g \mapsto \bar{g} := gZ$. This map allows to define a natural topology on G/Z: a subset V of G/Z is open if and only if $p^{-1}V$ is open in G. Note that if Z is a normal subgroup, then the quotient G/Z is also a topological group. In any case, for the determination of $\pi_1(G/Z)$ using $\pi_1(G)$ we have the following facts (see [117, §50, E]):

(1) $p_* : \pi_1(G) \to \pi_1(G/Z)$ is one-to-one.
(2) $p_*(\pi_1(G))$ is a normal subgroup of $\pi_1(G/Z)$.
(3) $\pi_1(G/Z)/p_*(\pi_1(G)) \simeq Z$.

As a consequence of these facts, we easily obtain the following additional facts:

(4) If G is simply connected, then $\pi_1(G/Z) \simeq Z$.
(5) If $Z \neq \{1\}$ and $\pi_1(G/Z) \simeq \mathbb{Z}_2$, then G is simply connected.

In turn, from these latter facts we get:

(6) Since $\mathrm{SO}_2 \simeq \mathrm{U}_1 \simeq \mathbb{R}/\mathbb{Z}$, $\pi_1(\mathrm{SO}_2) \simeq \mathbb{Z}$. Indeed, $(\mathbb{R}, +)$ is simply connected and $\mathbb{Z}$ is a discrete subgroup, so we can apply (4).
(7) Since $\mathrm{SO}_3 \simeq \mathcal{R}_3/\{\pm 1\}$ and $\mathcal{R}_3 \simeq S^3$, we have (again by (4))

$$\pi_1(\mathrm{SO}_3) \simeq \{\pm 1\} \simeq \mathbb{Z}_2.$$

(8) For all $n \geqslant 3$, $\pi_1(\mathrm{SO}_n) \simeq \mathbb{Z}_2$. This follows from (7), which is the case $n = 3$, and induction on n, for there is an isomorphism $\pi_1(\mathrm{SO}_n) \simeq \pi_1(\mathrm{SO}_{n-1})$. This last claim is deduced by using the action of SO_n on the sphere S^{n-1}. The key points are that the subgroup leaving invariant $\boldsymbol{e}_n$ is SO_{n-1}, so that we get a homeomorphism $\mathrm{SO}_n/\mathrm{SO}_{n-1} \simeq S^{n-1}$, and that S^{n-1} is simply connected (as $n - 1 \geqslant 2$). Indeed, with these facts the claim follows by fairly standard homotopy theory (see [62, 2nd ed., §13.2]).
(9) The group $\mathcal{R}_n = \mathcal{S}_n = \mathrm{Spin}_n$ is simply connected for $n \geqslant 3$ (use the isomorphism $\mathcal{R}_n/\{\pm 1\} \simeq \mathrm{SO}_n$, (8) and (4)). Note also that we have $\pi_1(\mathrm{Spin}_2) \simeq \pi_1(\mathrm{U}_1) \simeq \mathbb{Z}$.

For $r, s \geqslant 1$, it turns out that the group $\mathrm{O}_{r,s}$ is homeomorphic to $\mathrm{O}_r \times \mathrm{O}_s \times \mathbb{R}^d$, for some non-negative integer d. This can be deduced by means of polar decomposition techniques, like in [110, §4.2]. Note that this provides an alternative view of the fact that $O_{r,s}$ has four connected components (see 4.2.6 and 4.3.4). From this, and the facts explained so far, we get the following:

(10) $\pi_1(SO^0_{r,s}) = \pi_1(SO_r) \times \pi_1(SO_s)$. Indeed,

$$\pi_1(SO^0_{r,s}) = \pi_1(O_{r,s}) \simeq \pi_1(O_r \times O_s \times \mathbb{R}^d) \simeq \pi_1(O_r) \times \pi_1(O_s) \times \pi_1(\mathbb{R}^d)$$
$$\simeq \pi_1(SO_r) \times \pi_1(SO_s).$$

(11) $\pi_1(SO^0_{1,n}) \simeq \pi_1(SO^0_{n,1}) \simeq \pi_1(SO_n) = \begin{cases} \{0\} & \text{if } n = 1 \\ \mathbb{Z} & \text{if } n = 2 \\ \mathbb{Z}_2 & \text{if } n \geqslant 3. \end{cases}$

The result for $n = 1$ confirms what was found in 5.1.6, for there we saw that $SO^0_{1,1}$ is isomorphic to $(\mathbb{R}, +)$.

(12) $\pi_1(SO^0_{2,n}) \simeq \mathbb{Z} \times \pi_1(SO_n) = \begin{cases} \mathbb{Z} \times \mathbb{Z} & \text{if } n = 2 \\ \mathbb{Z} \times \mathbb{Z}_2 & \text{if } n \geqslant 3 \end{cases}$

and similarly $\pi_1(SO^0_{n,2}) \simeq \mathbb{Z}_2 \times \mathbb{Z}$ for $n \geqslant 3$.

(13) For $r, s \geqslant 3$, $\pi_1(SO^0_{r,s}) \simeq \mathbb{Z}_2 \times \mathbb{Z}_2$.

(14) Since $SO^0_{1,n} = \mathcal{R}_{1,n}/\{\pm 1\}$ (see Eq. (4.4)), (11) and (4) tell us that $\mathcal{R}_{1,n}$ is simply connected for $n \geqslant 3$. We will return to the case $\mathcal{R}_{1,3}$ in 6.5.11.

6.2 Grassmann's Heritage

Terms like *Grassmann's formula,*[7] *Grassmann's exterior algebra* (say as summarized in Chap. 2), and *Grassmannian manifolds* (▹ 6.5.4) are familiar concepts in mathematics books. The exterior algebra was developed by Poincaré and Cartan (cf. [138]) into the powerful formalism of exterior differential forms, widely used in analysis, geometry, and mathematical physics, and it is mostly in this form that Grassmann's algebra is known to mathematicians and to a good many theoretical physicists (see, for example, [22]).

But, they are often presented as abstract structures without delving into the innovative thinking that led Grassmann to their discovery. Abstract structures have a definite place both in mathematics and physics, but by themselves alone may leave the researcher wanting to know more about what philosophy of science and discovery launched them into existence. As voiced by Hestenes in [67], "in conception and applications, conventional renditions of his exterior algebra fall far short of Grassmann's original vision."

[7] If U and V are vector subspaces of a vector space, then

$$\dim(U + V) = \dim(U) + \dim(V) - \dim(U \cap V).$$

As it is often recognized, Grassmann was the first to advance the ideas of Leibniz on the *characteristica geometrica* (*geometric calculus* in today's terms). Actually, it can be defended that he initiated the language of mathematical structures which has developed as some incarnation of Leibniz' idea of *characteristica universalis* (cf. [148]). But to this day, there remain several puzzling circumstances in Grassmann's achievements. The best known may be that the two main mathematical works of Grassmann, namely Grassmann-1844 [58] and Grassmann-1862 [59] (see Grassmann-2000 [60] for L.C. Kannenberg's translation into English, with the title *Extension theory*, of the 1862 book), were not really appreciated in his time: "... despite writing a work which appears to us today to be in the style of a modern textbook, Grassmann failed to convince mathematicians of his own time."[8] One possible cause for this neglect may very well be that Grassmann's university training was in theology and linguistics, not in mathematics (see [116]), so the mathematical elites of his country were not inclined to value his achievements, and even less when they found them written in a rather unfamiliar language.

Neglected or not, the question still remains of what kindled Grassmann's mathematical mind. Due to Grassmann's education, one conjectural spark may have propagated from Leibniz' through philosophy (Kant, for example) and theology. In fact, according to [2], his core ideas about mathematical innovation came from Friedrich D.E. Schleiermacher (1768–1834), a philosopher, classical philologist, and theologian who wrote extensively on dialectics. More concretely, in [46] it is asserted that in 1831 Grassmann actually attended Schleiermacher's lectures on Leibniz' idea of a universal language. Again in [2], we read that those teachings "were picked up by Grassmann and operationalized in his philosophical-mathematical treatise Ausdehnungslehre in 1844."

Grassmann had great confidence in his theory (see the introduction to [60]), but it was not until Clifford combined it with Hamilton's quaternions [30], at around the end of Grassmann's life (1809–1877), that it began to bear additional and unsuspected fruits. Here, we can do no better than to refer the reader to Hestenes papers [67, 75], where the vision and impact of Grassmann's contributions are analyzed in depth.

6.3 The Development of GA

We have tried to reflect the view that GA is a mathematical structure that allows the representation of geometrical concepts, and operations on them, in an intrinsic manner, which essentially means that the formalism is coordinate-free. Its birth certificate is Clifford's discovery [30] of the *geometric product* and its capacity to fuse the ideas of Grassmann on extension calculus [58, 59] with those of Hamilton on quaternions [63]. Those ideas were rediscovered by Lipschitz [103, 104], who

8 http://www-history.mcs.st-andrews.ac.uk/Biographies/Grassmann.html.

also proposed and studied the versor groups that are rightly named after him (cf. [142]). Thereafter, in the twentieth century, the unfolding was rather slow, to a good extent caused by lack of communication between different interests and views.

6.3.1 (Timeline) Let us summarize the most salient advances, with brief comments when appropriate.

1908. Minkowski's space-time [109] distills the geometric content of special relativity. The Lorentz transformations are the isometries of a metric of signature $(1, 3)$.

1913. É. Cartan discovers the (infinitesimal) double-valued representations of the orthogonal group of n-dimensional Euclidean space [23]. About this work, in [24] Cartan says (from the English translation): "Spinors were first used under that name, by physicists, in the field of Quantum Mechanics. In their most general form, spinors were discovered in 1913 by the author of this work, in his investigations on the linear representations of simple groups; they provide a linear representation of the group of rotations in a space with any number n of dimensions, each spinor having 2^ν components where $n = 2\nu + 1$ or 2ν" (so two components for $n = 3$ and four for $n = 4$, which correspond to the Pauli and Dirac spinors coming next). Note that this notion of spinor (an element of a representation space) is quite different from the usage we have adopted in Sect. 4.3 (an element of the Spin group). At this point, the reader may like to consult [141, §6.1] (The Babel of Spinors).

1927. Pauli's theory of spin $\frac{1}{2}$ particles [113]. Rediscovers the GA of Euclidean space E_3 and Cartan's spinor representation (Pauli spinors) of SO_3.

1928. Dirac's relativistic equation for the spinning electron [35]. Rediscovers the GA of Minkowski's space-time $E_{1,3}$ and Cartan's spinor representation (Dirac spinors) of the Lorentz group ($SO^0_{1,3}$ with the notations of 5.2.3).
Weyl publishes [143] a landmark work laying the foundation for the mathematical treatment of symmetries in geometry and physics.

1935. Brauer and Weyl publish the theory spinors for n-dimensional Euclidean space [17]. This work extends Cartan's theory to representations (not necessarily infinitesimal) of the orthogonal group.

1937. E. Cartan's book on spinors [24].

1939. Weyl's treatise [144] on classical groups.

1954, 1955. Chevalley's algebraic theory of spinors [27, 28], reproduced in [29].

1957. E. Artin's *Geometric algebra* [6]. The title of this book may be misleading for all readers seeking knowledge about GA in the sense adopted in Chap. 3. One barrier may be its mathematical generality, particularly in working orthogonal and symplectic geometry over any field (Ch. III). Another barrier may be that Clifford algebra is not introduced until the very end of the text (§V.4), and then in a rather ad hoc formal manner ([27] is the main reference on this issue, but it is not followed in the definition via the tensor algebra). Nevertheless, in Definition 5.4 (of an even or odd *regular* element) we recognize the definition of Lipschitz versor with which we started 4.2, and Artin's Theorem 5.11

corresponds to the even part of 4.2.2, albeit for any field. Lipschitz stronger characterization of the versor group (as in E.4.2, p. 75) is not mentioned.

1958. Riesz [122]. This was an almost clandestine publication, and yet in [68] Hestenes qualifies it as the "midwife of the rebirth of geometric algebra." He tells the story as follows (emphasis not in the original): "I was consciously concerned with questions relating the structure of mathematical systems to the structure of the physical world. One day, Riesz's notes appeared on the new-book shelf of the UCLA library. The impact on me was immediate and striking! By the time, I was *half way through the first chapter I was convinced that Clifford algebra was the key to unifying mathematical physics*."

1964. Atiyah, Bott, and Shapiro publish *Clifford modules* [7]. This is an important mathematical paper that finds deep applications of Clifford algebras and Clifford modules (representations by another naming) to Topology.

1966. The Hestenes quotation from [68] (see 1958 above) continues as follows: "During the next few years I worked out the framework for a fully geometric unification. The result was published in my book *Space Time Algebra*" (STA) [65]. This was indeed a turning point, especially for geometrical physics, but later on, as explained in [68], many other applications were discovered. We will return to this in the next section, but let us add here that Hestenes' influence in the development of GA and its applications has been immense for over 50 years and that it continues to this day. For more details, we refer to the "Foreword" (A. Lasenby) and the "Preface after fifty years" included in the second edition (2015) of STA, and also to his Genesis paper [77]. Let us also say that Hestenes has increasingly favored using the term *Geometric Calculus* (GC) over the more restrictive GA: "*Geometric Calculus* is now sufficiently well-developed to serve as a comprehensive geometric language for the whole community of scientists, mathematicians and engineers" ([68], last sentence in the final Summary). In GC, the exterior differential d familiar to mathematicians is replaced by the vector (or Dirac) operator ∂ (cf. [98], §3.3). The great advantage is that ∂ behaves as a vector in the GA formalism and thereby enjoys all of its expressiveness. See Fig. 6.2.

1984, 1986. The bible of GC, [81]: "a self-contained system of mathematical tools sufficient for addressing any problem in physics without resorting to alternative mathematical formalisms" [68, p. 4]. See also [66].

1990, 1999. [69], Hestenes' *New foundations of Classical Mechanics*.

1993, 2001. Lounesto's *Clifford algebras and spinors* [105].

1995. Porteous, *Clifford algebras and the classical groups* [119].

1997, 2012. Snygg's books [128, 129].

2003. Doran and Lasenby, *Geometric algebra for physicists* [38].

2007. Dorst, Fontijne, and Mann' *Geometric algebra for computer scientists: an object-oriented approach to geometry* [42].

2007, 2016. Rodrigues and de Oliveira, *The many faces of Maxwell, Dirac, and Einstein's equations* [124].

2009. Perwass, *Geometric algebra with applications in engineering* [115].

2016. Vaz and da Rocha, *An introduction to Clifford algebras and spinors* [141].

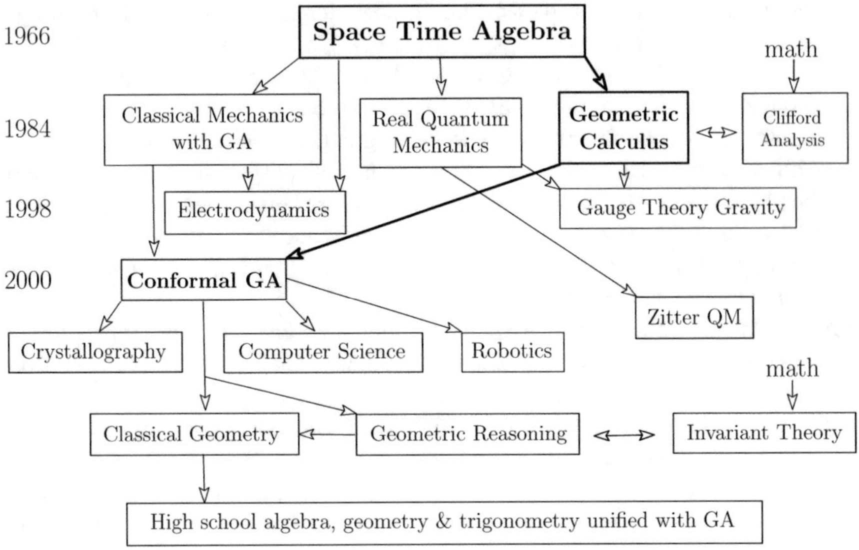

Fig. 6.2 The development of GC according to David Hestenes [75]

2018. Hestenes, *Maxwell-Dirac electron theory* [79], and *Quantum Mechanics of the electron particle-clock* [78].

6.3.2 (Clifford Algebra or Geometric Algebra?) Many authors take the view that GA is Clifford algebra (CA). For example, in the text [54], the term GA does not appear until the end of the last chapter and then it is used as an "alternative name" for CA. In an another excellent text [141], the term GA does not appear. On the other hand, texts such as [38] and [115] do not even mention Clifford Algebra.[9]

The issue is addressed at some length in §7.7.2 of [42], where even a *multiplicative principle* is advanced to distinguish between Clifford algebra and geometric algebra, which "we have found it useful for *practical purposes*, especially as a foundation for developing efficient implementations for the various admissible operations" (emphasis not in original).

The fact is, however, that the Clifford algebra $C_q(E) = C(E, q)$ is well defined for any orthogonal space (E, q) and that it has all the frills that we can hope for a geometric algebra *for any signature* (linear grading isomorphic to the linear grading of $\wedge E$, outer and inner products, and involutions). At one extreme, we have $q = 0$, whose Clifford algebra is just the Grassmann algebra $\wedge E$, and at the other we have the regular signatures which have been the focus of our attention in the preceding chapters (except for what is explained in Remark 3.2.1). The great difference in the non-regular case is that pseudoscalars are null and hence we do

[9]In fact, in Perwass' text Clifford only appears in the term "Clifford group," which actually corresponds to (a variation of) the Lipschitz group.

Table 6.1 For each integer ν mod 8, there is a basic algebra form that we denote by F_ν and which is given in the second row. The third row contains the dimensions $d_\nu = \dim F_\nu$. Now for any signature (r, s), we have $\mathcal{G}_{r,s} \simeq F_\nu(m)$, where $\nu = s - r \bmod 8$ and where m is determined from the relation $2^n = d_\nu m^2$ $(n = r + s)$. The fourth row applies the prescription to $\bar{\mathcal{G}}_\nu = \mathcal{G}_{0,\nu}$. For example, $\bar{G}_3 \simeq 2\mathbb{H}(m)$, with $2^3 = 8m^2$, hence $m = 1$ and $\bar{G}_3 \simeq 2\mathbb{H}$. Similarly, $\bar{\mathcal{G}}_5 \simeq \mathbb{C}(m)$, where $2^5 = 2 \times m^2$, and so $\bar{G}_5 \simeq \mathbb{C}(4)$. Note that $\bar{\mathcal{G}}_1 \simeq \mathbb{C}$ is immediate: if $\boldsymbol{e}$ is a unit vector of $\bar{E}_1 = E_{0,1}$, so that $\boldsymbol{e}^2 = -1$, then the isomorphism is given by $\alpha + \beta\boldsymbol{e} \mapsto \alpha + \beta i$, and the parity involution morphs into the complex conjugation

ν	0	1	2	3	4	5	6	7
F_ν	$\mathbb{R}$	$\mathbb{C}$	$\mathbb{H}$	$2\mathbb{H}$	$\mathbb{H}$	$\mathbb{C}$	$\mathbb{R}$	$2\mathbb{R}$
d_ν	1	2	4	8	4	2	1	2
$\bar{\mathcal{G}}_\nu$	$\mathbb{R}$	$\mathbb{C}$	$\mathbb{H}$	$2\mathbb{H}$	$\mathbb{H}(2)$	$\mathbb{C}(4)$	$\mathbb{R}(8)$	$2\mathbb{R}(8)$

Table 6.2 For n in $0, 1, \ldots, 8$, the second row displays $\nu = 0 - n \bmod 8$ and the third the isomorphism class of $\mathcal{G}_n = \mathcal{G}_{n,0}$. For example, $\mathcal{G}_6 \simeq F_2(m) = \mathbb{H}(m)$, where $2^6 = 4m^2$, and hence $\mathcal{G}_6 \simeq \mathbb{H}(4)$

n	0	1	2	3	4	5	6	7	8
ν	0	7	6	5	4	3	2	1	0
$\mathcal{G}_n$	$\mathbb{R}$	$2\mathbb{R}$	$\mathbb{R}(2)$	$\mathbb{C}(2)$	$\mathbb{H}(2)$	$2\mathbb{H}(2)$	$\mathbb{H}(4)$	$\mathbb{C}(8)$	$\mathbb{R}(16)$

not have Hodge duality. So, it may be a pragmatic idea to reserve the name GA for regular metrics and the term CA for arbitrary metrics. The bottom line is that CA provides a powerful formalism which by its very nature carries a rich semantics, be it for Clifford Analysis in general, as in [89], or for geometry in the regular case. Nobody should consider these terminology questions as thought-limiting in any way.

Notations In what follows, we will let $\mathbb{K}$ be one of the fields $\mathbb{R}, \mathbb{C}, \mathbb{H}$. A $\mathbb{K}$-vector space is defined by mimicking the definition of a (real) vector space at the beginning of Sect. 1.2, and we will speak of *real*, *complex*, or *quaternionic* (or *symplectic*) vector spaces when $\mathbb{K}$ is $\mathbb{R}$, $\mathbb{C}$, or $\mathbb{H}$, respectively. Since $\mathbb{R}$ and $\mathbb{C}$ are commutative, for real and complex vector spaces we have the freedom to write the product of a scalar and a vector in whatever order we please. But for quaternionic vector spaces, this freedom disappears because one has to distinguish between *left* and *right* quaternionic vector spaces. In this text, we will only consider right quaternionic spaces, which means that the product of a vector $\boldsymbol{v}$ times a quaternion $h \in \mathbb{H}$ is always written as $\boldsymbol{v}h$.

6.3.3 (The Clifford ν Clock) The isomorphism class of the geometric algebras $\mathcal{G}_{r,s}$ can be explicitly specified in terms of $\nu = s - r \bmod 8$ according to the prescriptions explained in Table 6.1.

Note that for $1 \leqslant n \leqslant 8$ we have $\mathcal{G}_n = \mathcal{G}_{n,0} \simeq F_{8-n}(m)$, for some m that depends on n. This immediately yields, setting $\nu = 8 - n \bmod 8$, the data in Table 6.2. Another remark is that E.3.8, p. 60 tells us that the even algebra $\mathcal{G}^+_{r,s}$ is isomorphic to $F_{\nu-1}(m^+)$, where m^+ is determined by the relation $2^{n-1} = (m^+)^2 d_{\nu-1}$.

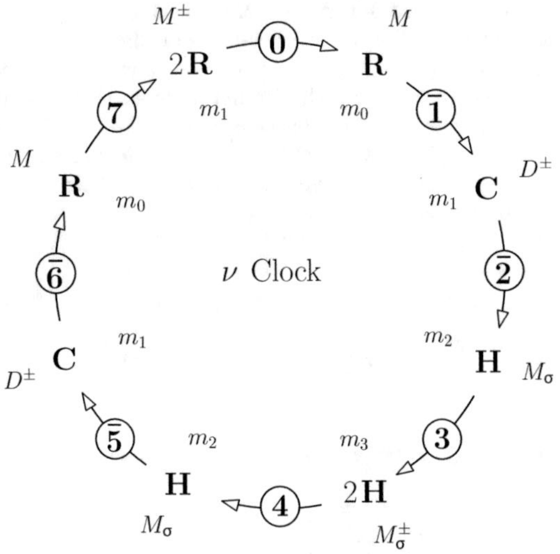

Fig. 6.3 Ahead of the arrow of any hour ν, we have the form F_ν of $\mathcal{G}_{r,s}$, where $\nu = s - r \bmod 8$. Therefore, the form $F_{\nu-1}$ of $\mathcal{G}^+_{r,s}$ can be read at the tail of the ν-arrow. To specify the order m of the matrices, it is convenient to use the notation $m_k = 2^{(n-k)/2}$. Here, $k = 0, \ldots, 3$, but later we will also need m_4. For example, $\mathcal{G}_{3,1} \simeq F_6(m) = \mathbb{R}(m_0)$, where $m_0 = 2^{4/2} = 4$, which tells us that $\mathcal{G}_{3,1}$ is isomorphic, as an algebra, to the matrix algebra $\mathbb{R}(4)$. On the other hand, $\mathcal{G}^+_{3,1}$ is isomorphic to $\mathbb{C}(m_1) = \mathbb{C}(2)$, because $\nu = 6$, $F_5 = \mathbb{C}$, and $m_1 = 2^{(3-1)/2} = 2$. The values $\nu = 1, 2, 5, 6$ (or $\nu = 1, 2 \bmod 4$) have been marked with an overbar and its significance is explained in 6.3.5. The labels M and M_σ stand for *real* and *symplectic* (or quaternionic) *Majorana*, respectively, and D for *Dirac*, and their significance is explained in 6.3.6–6.3.8

All the classification results, and others to come, are best summarized in the 8-h *ν-clock* (cf. Budinich-Trautman-1988 [19]) shown in Fig. 6.3.

6.3.4 (Remarks About the Proof) We already know that $\mathcal{G}_0 = \mathbb{R}$ (obvious), $\mathcal{G}_1 \simeq 2\mathbb{R}$ (E.3.4, p. 58), $\mathcal{G}_2 \simeq \mathbb{R}(2)$ (E.3.5, p. 58), and $\mathcal{G}_3 \simeq \mathbb{C}(2)$ (E.3.7, p. 60).

It is also easy to see $\bar{\mathcal{G}}_2 \simeq \mathbb{H}$ (see Table 6.1): if $\boldsymbol{e}_1, \boldsymbol{e}_2$ is an orthonormal basis of $\bar{E}_2 = E_{0,2}$, then the isomorphism is given by:

$$\alpha + \beta_1 \boldsymbol{e}_1 + \beta_2 \boldsymbol{e}_2 + \beta_{12} \boldsymbol{e}_1 \boldsymbol{e}_2 \mapsto \alpha + \beta_1 \boldsymbol{i} + \beta_2 \boldsymbol{j} + \beta_{12} \boldsymbol{k}.$$

This is because $\boldsymbol{e}_1, \boldsymbol{e}_2, \boldsymbol{e}_1\boldsymbol{e}_2$ satisfy Hamilton's relations, as for example:

$$\boldsymbol{e}_1^2 = \boldsymbol{e}_2^2 = (\boldsymbol{e}_1\boldsymbol{e}_2)^2 = -1 \text{ and } \boldsymbol{e}_1(\boldsymbol{e}_1\boldsymbol{e}_2) = -\boldsymbol{e}_2 = -(\boldsymbol{e}_1\boldsymbol{e}_2)\boldsymbol{e}_1.$$

In this case, the quaternion conjugation is given by the *Clifford conjugation*, that is, the composition of the parity and reverse involutions.

To go further, we need two key isomorphisms: for any $n \geqslant 0$, we have

$$\mathcal{G}_{n+2} \simeq \bar{\mathcal{G}}_n \otimes \mathcal{G}_2 \simeq \bar{\mathcal{G}}_n(2), \quad \bar{\mathcal{G}}_{n+2} \simeq \mathcal{G}_n \otimes \bar{\mathcal{G}}_2 \simeq \mathcal{G}_n \otimes \mathbb{H}. \tag{6.3}$$

To see this, consider E_{n+2} as $E_n \perp E_2$, and let $\boldsymbol{u}$, $\boldsymbol{v}$ be an orthonormal basis of E_2. Then, consider the linear map $f : E_{n+2} \to \bar{\mathcal{G}}_n \otimes \mathcal{G}_2$ given by:

$$\boldsymbol{x} + \lambda\boldsymbol{u} + \mu\boldsymbol{v} \mapsto \bar{\boldsymbol{x}} \otimes \boldsymbol{uv} + \lambda(1 \otimes \boldsymbol{u}) + \mu(1 \otimes \boldsymbol{v}),$$

where $\bar{\boldsymbol{x}} \in \bar{E}_n$ is the vector $\boldsymbol{x} \in E_n$ but with $\bar{\boldsymbol{x}}^2 = -\boldsymbol{x}^2$. With this, it is immediate to check that $f(\boldsymbol{x}+\lambda\boldsymbol{u}+\mu\boldsymbol{v})^2 = (\boldsymbol{x}^2+\lambda^2+\mu^2)(1\otimes 1) = q(\boldsymbol{x}+\lambda\boldsymbol{u}+\mu\boldsymbol{v})(1\otimes 1)$ and hence f extends uniquely to an algebra homomorphism $\bar{f} : \mathcal{G}_{n+2} \to \bar{\mathcal{G}}_n \otimes \mathcal{G}_2$. Since any element of $\mathcal{G}_{n+2}$ can be written in a unique way in the form $a + b\boldsymbol{u} + c\boldsymbol{v} + d\boldsymbol{uv}$, with $a, b, c, d \in \mathcal{G}_n$, we have $\bar{f}(a+b\boldsymbol{u}+c\boldsymbol{v}+d\boldsymbol{uv}) = \bar{a}\otimes 1+\bar{b}\otimes\boldsymbol{u}+\bar{c}\otimes\boldsymbol{v}+\bar{d}\otimes\boldsymbol{uv}$, where $\bar{a} \in \bar{\mathcal{G}}_n$, say, is the same as $a \in \mathcal{G}_n$ but obeying the metric of $\bar{E}_n$ instead of that of E_n. It is thus clear that $\bar{f}$ is a linear isomorphism and hence an algebra isomorphism. The other isomorphism in (6.3) can be proved in an analogous way.

Now using (6.3), we immediately get

$$\bar{\mathcal{G}}_3 \simeq \mathcal{G}_1 \otimes \mathbb{H} \simeq (2\mathbb{R}) \otimes \mathbb{H} \simeq 2\mathbb{H} \tag{6.4}$$

$$\mathcal{G}_4 \simeq \bar{\mathcal{G}}_2(2) \simeq \mathbb{H}(2) \tag{6.5}$$

$$\bar{\mathcal{G}}_4 \simeq \mathcal{G}_2 \otimes \mathbb{H} \simeq \mathbb{R}(2) \otimes \mathbb{H} \simeq \mathbb{H}(2) \tag{6.6}$$

$$\mathcal{G}_5 \simeq \bar{\mathcal{G}}_3(2) \simeq (2\mathbb{H})(2) \simeq 2\mathbb{H}(2) \tag{6.7}$$

$$\bar{\mathcal{G}}_5 \simeq \mathcal{G}_3 \otimes \mathbb{H} \simeq \mathbb{C}(2) \otimes \mathbb{H} \simeq \mathbb{C}(4) \tag{6.8}$$

as $\mathbb{C} \otimes \mathbb{H} \simeq \mathbb{C}(2)$ (see E.5.1, p. 103)

$$\mathcal{G}_6 \simeq \bar{\mathcal{G}}_4(2) \simeq \mathbb{H}(4) \tag{6.9}$$

$$\bar{\mathcal{G}}_m \simeq \mathcal{G}_4 \otimes \mathbb{H} \simeq (\mathbb{H}(2)) \otimes \mathbb{H} \simeq \mathbb{R}(8) \tag{6.10}$$

as $\mathbb{H} \otimes \mathbb{H} \simeq \mathbb{R}(4)$ (see E.5.2, p. 103)

$$\mathcal{G}_7 \simeq \bar{\mathcal{G}}_5(2) \simeq \mathbb{C}(8) \tag{6.11}$$

$$\bar{\mathcal{G}}_7 \simeq \mathcal{G}_5 \otimes \mathbb{H} \simeq (2\mathbb{H}(2)) \otimes \mathbb{H} \simeq 2\mathbb{R}(8) \tag{6.12}$$

So far, we have established 6.3.3 for signatures $(n, 0)$ and $(0, n)$ with $0 \leqslant n \leqslant 7$. If $n \geqslant 8$, then repeating the isomorphisms (6.3) four times we conclude that

$$\mathcal{G}_n \simeq \mathcal{G}_{n-8} \otimes \bar{\mathcal{G}}_2 \otimes \mathcal{G}_2 \otimes \bar{\mathcal{G}}_2 \otimes \mathcal{G}_2 \simeq \mathcal{G}_{n-8}(16), \tag{6.13}$$

because $\bar{G}_2 \otimes \mathcal{G}_2 \otimes \bar{G}_2 \otimes \mathcal{G}_2 \simeq \mathbb{H}\otimes\mathbb{R}(2)\otimes\mathbb{H}\otimes\mathbb{R}(2) \simeq \mathbb{H}(2)\otimes\mathbb{H}(2) \simeq \mathbb{R}(16)$. This proves the assertion for all signatures $(n, 0)$. For signatures $(0, n)$, the argument is similar, for we get

$$\bar{\mathcal{G}}_n \simeq \bar{\mathcal{G}}_{n-8} \otimes \mathcal{G}_2 \otimes \bar{\mathcal{G}}_2 \otimes \mathcal{G}_2 \otimes \bar{\mathcal{G}}_2 \simeq \bar{\mathcal{G}}_{n-8}(16). \tag{6.14}$$

Table 6.3 Synopsis of the $\mathcal{G}_{r,s}$ forms in terms of $\nu = s - r \bmod 8$. By definition, $m_k = 2^{(n-k)/2}$, $n = r + s$, is the dimension of the ground space and the order of the matrices in the algebra class. In the labels, M and D stand for Majorana and Dirac, respectively

ν	0, 6	2, 4	1, 5	7	3
F_ν	$\mathbb{R}$	$\mathbb{H}$	$\mathbb{C}$	$2\mathbb{R}$	$2\mathbb{H}$
m	m_0	m_2	m_1	m_1	m_3
Label	M	M_σ	$D^\pm$	$M^\pm$	$M_\sigma^\pm$

Finally, for signatures (r, s) with $r, s > 0$, by E.3.6, p. 58 we have $\mathcal{G}_{r,s} \simeq \mathcal{G}_{r-1,s-1}(2)$, which by induction has the form $F_\nu(2m)$ because $\nu = s - r = (s-1) - (r-1) \bmod 8$.

Equations (6.13) and (6.14) display what is called 8-*periodicity* of the isomorphism class of the geometric algebras $\mathcal{G}_{r,s}$ in terms of ν.

6.3.5 (The Key Role of the Pseudoscalar in the Classification) We will write $I = I_{r,s}$ to denote a unit pseudoscalar of $\mathcal{G}_{r,s}$. It is defined up to sign and we easily get

$$I^2 = \begin{cases} 1 & \text{if } \nu = 0, 3, 4, 7 \quad (0, 3 \bmod 4), \\ -1 & \text{if } \nu = 1, 2, 5, 6 \quad (1, 2 \bmod 4). \end{cases} \tag{6.15}$$

In fact, we have $\nu = s - r \pmod 8$, so that $s = r + \nu \pmod 8$, $n = r + s = 2r + \nu \pmod 8$, and by 3.4.1(1) the sign of I^2 coincides with the parity of

$$n /\!\!/ 2 + s = r + \nu /\!\!/ 2 + r + \nu \bmod 8,$$

which is the parity of $(3\nu)/\!\!/2$, or more simply of $(\nu + 1)/\!\!/2$.

Recall also that for any vector $\boldsymbol{x}$, $\boldsymbol{x}I = (-1)^{n-1} I\boldsymbol{x}$, so that I is central for n odd and anticommutes with vectors if n is even (so it anticommutes with odd multivectors and commutes with even multivectors). Since $n \equiv \nu \bmod 2$, I is central if and only if ν is odd.

The form corresponding to each of these possibilities is explained in the next three blocks and summarized in Table 6.3.

6.3.6 (Simple Real and Quaternionic Algebras (ν Even)) The form for the case $\nu = 0$, or $r = s = n/2$, is $\mathbb{R}(m_0)$, as seen by iterating $n/2$ times the isomorphism obtained in E.3.6, p. 58. The ground space $\mathbb{R}^{m_0}$ on which $\mathbb{R}(m_0)$ acts is a *real Majorana space* (symbol M in the ν-clock).

For $\nu = 2$ and $\nu = 4$, we get the isomorphism class $\mathbb{H}(m_2)$. Indeed, this holds for $n = 2$, as $\bar{G}_2 \simeq \mathbb{H}$ and $m_2 = 1$ in this case, and for $\nu = 4$, as $\bar{G}_4 \simeq \mathbb{H}(2)$ and $m_2 = 2$ in this case. Now by the 8-periodicity, we get

$$\bar{\mathcal{G}}_{4+8k} \simeq \mathbb{H}(2 \times 2^{4k}) = \mathbb{H}(m_2),$$

for $n = 4 + 8k$ and $4k + 1 = (n-2)/2$. The ground space $\mathbb{H}^{m_2}$ on which $\mathbb{H}(m_2)$ acts is a *symplectic Majorana space* (symbol M_σ).

It remains the case $\nu = 6$, whose isomorphism class is $\mathbb{R}(m_0)$, with real Majorana ground space $M = \mathbb{R}^{m_0}$ as for $\nu = 0$. As in the above cases, this follows from the isomorphism $\bar{\mathcal{G}}_6 \simeq \mathbb{R}(4)$ and the 8-periodicity.

6.3.7 (The Split Cases $\nu = 3, 7$) These cases correspond to n odd (so I is central) and $I^2 = 1$, which express (see 3.2.10) that the signature is special. Now by E.3.9, p. 61 we know that

$$\mathcal{G} \simeq 2\mathcal{G}^+ = 2F_{\nu-1}(m) = \begin{cases} 2\mathbb{H}(m_3) & \text{if } \nu = 3, \\ 2\mathbb{R}(m_1) & \text{if } \nu = 7. \end{cases}$$

There are two quaternionic ground spaces $\mathbb{H}^{m_3}$ on which $\mathbb{H}(m_3)$ acts which are distinguished by what factor we choose of the two in $2\mathbb{H}(m_3)$, and ultimately this is reflected in the action ± 1 of I on those factors. These two spaces are *symplectic Majorana spaces* and labeled by $M_\sigma^\pm$. Similarly, there are two real ground spaces $\mathbb{R}^{m_1}$ on which $\mathbb{R}(m_1)$ acts which are *real Majorana spaces* and are labeled $M^\pm$.

6.3.8 (The Complex Cases, $\nu = 1, 5$) The pseudoscalar I still commutes with vectors, but $I^2 = -1$. If we let $\mathbf{C} = \langle 1, I\rangle \simeq \mathbb{C}$, then $\mathcal{G}$ is a $\mathbf{C}$-algebra. Since $I\mathcal{G}^+ = \mathcal{G}^-$, we see that $\mathcal{G} = \mathcal{G}^+ + I\mathcal{G}^+ = \mathbf{C}\mathcal{G}^+$, which implies that the natural algebra homomorphism $\mathbf{C} \otimes \mathcal{G}^+ \to \mathcal{G}$ (determined by $I \otimes x \mapsto Ix$) is an isomorphism. Since

$$\mathcal{G}^+ \simeq \bar{G}_{\nu-1}(m_1) \simeq \begin{cases} \mathbb{R}(m_1) & \text{if } \nu = 1, \\ \mathbb{H}(m_3) & \text{if } \nu = 5, \end{cases}$$

we finally get $\mathcal{G} \simeq \mathbb{C}(m_1)$ for $\nu = 1$ and $\mathcal{G} \simeq (\mathbb{C}\otimes\mathbb{H})(m_3)$ for $\nu = 5$. But, $\mathbb{C}\otimes\mathbb{H} \simeq \mathbb{C}(2)$ and so $\mathcal{G} \simeq \mathbb{C}(2m_3) = \mathbb{C}(m_1)$ as well. The ground space $\mathbb{C}^{m_1}$ on which $\mathbb{C}(m_1)$ acts is called the *Dirac space*. On changing the orientation of our space E, I is replaced by $-I$, and the corresponding Dirac space becomes the complex conjugate of D. We label as $D^\pm$ this pair of complex conjugate spaces.

6.3.9 (Generalities on Representations) A *representation* of a (real) algebra $\mathcal{A}$ in a $\mathbb{K}$-vector space V is an algebra homomorphism $\rho : \mathcal{A} \to \mathrm{End}_{\mathbb{K}}(V)$, $a \mapsto \rho_a$, where $\mathrm{End}_{\mathbb{K}}(V)$ is regarded as an $\mathbb{R}$-algebra. Depending on $\mathbb{K}$, we speak of *real*, *complex*, or *quaternionic* representations of $\mathcal{A}$.

A $\mathbb{K}$-vector subspace W of V is said to be *ρ-invariant* if it is invariant by all the endomorphisms ρ_a, $a \in \mathcal{A}$. If only the ρ-invariant subspaces are trivial (0 or V), the representation is said to be *irreducible*.

Given another $\mathbb{K}$-representation $\rho' : \mathcal{A} \to \mathrm{End}_{\mathbb{K}}(V')$, we say that it is equivalent to ρ if there is a $\mathbb{K}$-linear isomorphism $\varphi : V \to V'$ such that $\varphi\circ\rho_a = \rho'_a\circ\varphi$.

Similar definitions can be phrased for groups G, instead of algebras $\mathcal{A}$, by replacing $\mathrm{End}_{\mathbb{K}}(V)$ by $\mathrm{GL}_{\mathbb{K}}(V)$ and by requiring that ρ be a group homomorphism.

6.3.10 (Representations of Matrix Algebras) We are going to use the following algebraic fact (see [3]):

(1) Every irreducible real representation of $\mathbb{R}(n)$ is isomorphic to the standard representation in $\mathbb{R}^n$.
(2) Every irreducible real representation of $\mathbb{H}(n)$ is isomorphic to the standard representation in $\mathbb{H}^n$.
(3) Every irreducible complex representation of $\mathbb{C}(n)$ is isomorphic either to $\mathbb{C}^n$ or to $\bar{\mathbb{C}}^n$. The latter is defined so that a complex matrix Z acts as the multiplication by $\bar{Z}$.
(4) For $\mathbb{K} = \mathbb{R}$ or $\mathbb{K} = \mathbb{H}$, any real irreducible representation of $\mathbb{K}(m) \times \mathbb{K}(m)$ is isomorphic to one of the two standard representations in $\mathbb{K}^m$.

6.3.11 (Pinor and Spinor Representations) A *pinor representation* of $\mathrm{Pin}_{r,s}$ is the restriction to $\mathrm{Pin}_{r,s}$ of an *irreducible* representation of $\mathcal{G}_{r,s}$. The type of the pinor representations depends only on ν (and n) and they can be determined, using the spinorial clock, by the following algorithm:

$$\nu \begin{cases} \text{even (unique form } P) \begin{cases} 0, 6 \to \mathbb{R}^{m_0}\ (M) \\ 2, 4 \to \mathbb{H}^{m_2}\ (M_\sigma) \end{cases} \\ \\ \text{odd (two forms } P, \bar{P}) \begin{cases} 1, 5\ (\text{so } I^2 = -1) \to \mathbb{C}^{m_1}, \bar{\mathbb{C}}^{m_1}\ (D^\pm) \\ 3, 7\ (\text{so } I^2 = 1) \begin{cases} 7 \to \mathbb{R}^{m_1}, \bar{\mathbb{R}}^{m_1}\ (M^\pm) \\ 3 \to \mathbb{H}^{m_3}, \bar{\mathbb{H}}^{m_3}\ (M_\sigma^\pm) \end{cases} \end{cases} \end{cases}$$

These results have been reflected in Fig. 6.3 so that it can be used to find the pinor representation corresponding to n and ν at a glance.

A *spinor representation* of $\mathrm{Spin}_{r,s}$ is the restriction to $\mathrm{Spin}_{r,s}$ of an *irreducible* representation of $\mathcal{G}^+_{r,s}$. By looking at the tail of the ν arrows in the ν-clock, and with a similar argument as for pinor representations, the spinor representations can be classified with the following procedure (the label W stands for *Weyl*):

$$\nu \begin{cases} \text{odd (unique form } S) \to \begin{cases} 1, 7 \to \mathbb{R}^{m_1}\ (M) \\ 3, 5 \to \mathbb{H}^{m_3}\ (M_\sigma) \end{cases} \\ \\ \text{even (two forms } S, \bar{S}) \begin{cases} 2, 6\ (\text{so } I^2 = -1) \to \mathbb{C}^{m_2}, \bar{\mathbb{C}}^{m_2}\ (WD^\pm) \\ 0, 4\ (\text{so } I^2 = 1) \to \begin{cases} 0 \to \mathbb{R}^{m_2}, \bar{\mathbb{R}}^{m_2}\ (WM) \\ 4 \to \mathbb{H}^{m_4}, \bar{\mathbb{H}}^{m_4}\ (WM_\sigma) \end{cases} \end{cases} \end{cases}$$

Notice that the forms for $\nu = 5, 6, 7$ are the same as those for $\nu = 3, 2, 1$. This means that the row of the eight forms indexed by $\nu = 0, \ldots, 7$ is symmetric with respect to $\nu = 4$. Note also that in each case instead of m_k we have m_{k+1} because the dimension of the even algebra is half the dimension of the full algebra (we have to replace n by $n-1$ and so the matrix order m is $2^{(n-1-k)/2} = m_{k+1}$).

Table 6.4 Pinor and spinor representations for $1 \leqslant n \leqslant 7$

(r,s)	(1,0)	(0,1)
ν	7	1
$P, \bar{P}$	$\mathbb{R}, \bar{\mathbb{R}}$	$\mathbb{C}, \bar{\mathbb{C}}$
S	$\mathbb{R}$	$\mathbb{R}$

(r,s)	(2,0)	(1,1)	(0,2)
ν	6	0	2
P	$\mathbb{R}^2$	$\mathbb{R}^2$	$\mathbb{H}$
$S, \bar{S}$	$\mathbb{C}, \bar{\mathbb{C}}$	$\mathbb{R}, \bar{\mathbb{R}}$	$\mathbb{C}, \bar{\mathbb{C}}$

(r,s)	(3,0)	(2,1)	(1,2)	(0,3)
ν	5	7	1	3
$P, \bar{P}$	$\mathbb{C}^2, \bar{\mathbb{C}}^2$	$\mathbb{R}^2, \bar{\mathbb{R}}^2$	$\mathbb{C}^2, \bar{\mathbb{C}}^2$	$\mathbb{H}, \bar{\mathbb{H}}$
S	$\mathbb{H}$	$\mathbb{R}^2$	$\mathbb{R}^2$	$\mathbb{H}$

(r,s)	(4,0)	(3,1)	(2,2)	(1,3)
ν	4	6	0	2
P	$\mathbb{H}^2$	$\mathbb{R}^4$	$\mathbb{R}^4$	$\mathbb{H}^2$
$S, \bar{S}$	$\mathbb{H}, \bar{\mathbb{H}}$	$\mathbb{C}^2, \bar{\mathbb{C}}^2$	$\mathbb{R}^2, \bar{\mathbb{R}}^2$	$\mathbb{C}^2, \bar{\mathbb{C}}^2$

(r,s)	(5,0)	(4,1)	(3,2)	(2,3)
ν	3	5	7	1
$P, \bar{P}$	$\mathbb{H}^2, \bar{\mathbb{H}}^2$	$\mathbb{C}^4, \bar{\mathbb{C}}^4$	$\mathbb{R}^4, \bar{\mathbb{R}}^4$	$\mathbb{C}^4, \bar{\mathbb{C}}^4$
S	$\mathbb{H}^2$	$\mathbb{H}^2$	$\mathbb{R}^4$	$\mathbb{R}^4$

(r,s)	(6,0)	(5,1)	(4,2)	(3,3)
ν	2	4	6	0
P	$\mathbb{H}^4$	$\mathbb{H}^4$	$\mathbb{R}^8$	$\mathbb{R}^8$
$S, \bar{S}$	$\mathbb{C}^4, \bar{\mathbb{C}}^4$	$\mathbb{H}^2, \bar{\mathbb{H}}^2$	$\mathbb{C}^4, \bar{\mathbb{C}}^4$	$\mathbb{R}^4, \bar{\mathbb{R}}^4$

(r,s)	(7,0)	(6,1)	(5,2)	(4,3)
ν	1	3	5	7
$P, \bar{P}$	$\mathbb{C}^8, \bar{\mathbb{C}}^8$	$\mathbb{H}^4, \bar{\mathbb{H}}^4$	$\mathbb{C}^8, \bar{\mathbb{C}}^8$	$\mathbb{R}^8, \bar{\mathbb{R}}^8$
S	$\mathbb{R}^8$	$\mathbb{H}^4$	$\mathbb{H}^4$	$\mathbb{R}^8$

In each case, the first row contains up to the first four $n+1$ signatures, the second the corresponding ν's, and the third and fourth the pinor and spinor representations. For $n = 2$, the action of $\mathrm{Spin}_2 = \mathrm{U}_1$ on the representation spaces $\mathbb{C}$ and $\bar{\mathbb{C}}$ is by multiplication and conjugate multiplication. To note also that the spinor representation for $\mathcal{G}_3$ is $\mathbb{H}$ and that $\mathrm{Spin}_3 = \mathrm{SU}_2$ acts on this space by left multiplication. Similarly, the spin representations of $\mathcal{G}_{1,3}$ are $\mathbb{C}^2$ (*Pauli representation*) and $\bar{\mathbb{C}}^2$ (Pauli conjugate representation); the sum representation $\mathbb{C}^2 \otimes \bar{\mathbb{C}}^2$ is the *Dirac representation*

For a given n ($1 \leqslant n \leqslant 7$), there are $n+1$ signatures: $(r, n-r)$, $0 \leqslant r \leqslant n$. The corresponding $\nu = 2r - n$ decrease from $\nu = n$ to $\nu = -n$ in steps of -2, but in case $n \geqslant 3$ it is only necessary to find the forms for the first four values of ν because the remaining $n - 3$ cases repeat the beginning of the sequence, as $\nu(r,s) = \nu(r+4, s-4) \bmod 8$ (Table 6.4).

The topics dealt with in this section can be found, in one form or another, in many excellent texts. Here is a sample of those that I could consult, ordered by the year of first publication: [27, 33, 44, 54, 99, 105, 120, 135, 140, 141].

6.4 Applications

One strong message the timeline in Sect. 6.3 highlights is that the main driving force in the development of GA and GC has been the need to optimize a formalism in view of applications to mathematical physics, from forerunners like [17, 23, 24, 35, 109, 113, 143, 144], and [122] to the era ushered by Hestenes [65], reinforced in works like [66, 69, 72, 73], and continued by works such as [38, 124, 141], and [78, 79]. But, we also saw that in the last 20 years there has been an explosion of other applications (cf. Fig. 6.2), again initiated by several advances in which Hestenes played a pioneering role: [82] on projective geometry; [83, 84, 102], and [101], which initiate, among many other goods, Conformal Geometric Algebra (CGA), for which we also refer to the treatises [42] and [115]; [70] and [74] on approaches to computational geometry; [80] on crystallography; and still others that we will find below.

6.4.1 (Witnesses) Here are some prominent witnesses of that explosion:

- [130]. Features contributions of 20 researchers, grouped in three broad areas: A Unified Algebraic Approach for Classical Geometry, 6 papers (in the first four, already cited in the previous paragraph, Hestenes is one of the coauthors); Algebraic Embedding of Signal Theory and Neural Computation, 7 papers; Geometric Algebra for Computer Vision and Robotics, 8 papers.
- [13]. With contributions of 33 researchers, grouped in seven areas: Advances in GA, which includes Hestenes presentation of CGA [70]; Theorem Proving; Computer Vision; Robotics; Quantum and Neural Computing, and Wavelets; Applications to Engineering and Physics; and Computational Methods in Clifford Algebras.
- [41]. With contributions of 60 researchers, grouped in four broad areas: Algebra and Geometry, 18 papers, [71] among them; Applications to Physics, 9 papers; Computer Vision and Robotics, 10 papers; and Signal Processing and Other Applications, 5 papers.
- [10]. With contributions of some 40 researchers, grouped in the following areas: Neuroscience; Neural Networks; Image Processing; Computer Vision; Perception and Action; Uncertainty in Geometric Computations; Computer Graphics and Visualization; Geometry and Robotics; and Reaching and Motion Planning.
- [11]. It is worth taking notice of the six parts in which this 625-page single author treatise is divided: Fundamentals of Geometric Algebra; Euclidean, Pseudo-Euclidean, Lie and Incidence Algebras and Conformal Geometries; Geometric Computing for Image Processing, Computer Vision, and Neurocomputing; Geometric Computing of Robot Kinematics and Dynamics; Applications I: Image Processing, Computer Vision, and Neurocomputing; and Applications II: Robotics and Medical Robotics. The volume has a Foreword by D. Hestenes of which we extract the following appreciation: "This book assembles diverse aspects of geometric algebra in one place to serve as a general reference for applications to robotics. Then, it demonstrates the power and efficiency of the system with specific applications to a host of problems ranging from computer vision to mechanical control. Perceptive readers will recognize many places where the treatment can be extended or improved. Thus, this book is a work in progress, and its higher purpose will be served if it stimulates further research and development."
- [12]. With contributions of 50 researchers, grouped in seven areas: Geometric Algebra (4 chapters, starting with [74]); Clifford Fourier Transform (5 chapters); Image processing, Wavelets and Neurocomputing (4 chapters); Computer Vision (2 chapters); Conformal mapping and fluid analysis (2 chapters); Crystallography, Holography and Complexity (3 chapters); and Efficient computing with Clifford (Geometric) Algebra (3 chapters).

- [40]. With contributions of 34 researchers, grouped in eight areas: Rigid Body Motion, four papers; Interpolation and Tracking, three papers; Image Processing, two papers; Theorem Proving and Combinatorics, three papers; Applications of Line Geometry, two papers; Alternatives to Conformal Geometric Algebra, four papers; Towards Coordinate-Free Differential Geometry, two papers, including [76]; and a Tutorial Appendix [39].

The situation is that GA has been applied or connected to a great variety of fields (see the synopsis assembled in [96] for a general charter up to 2011), including mechanics, electromagnetism and wave propagation, general relativity, cosmology, robotics, computer graphics, computer vision, molecular geometry, symbolic algebra, automated theorem proving, and quantum computing, and its bearing on deep learning seems a promising idea.

6.4.2 (Other References) Spinors in physics: [15, 87]; Electrodynamics: [8, 90]; CGA and geometric computing: [9, 31, 32, 85]; Oriented projective geometry and oriented CGA: [21, 134]; Distance geometry: [64, 98]; Applications to Physics and Engineering: [1, 11, 52]; and Symbolic geometric reasoning: [100]

6.4.3 (Projective and Conformal Geometry) With geometric algebra, the group of conformal transformations of a space $E = E_{r,s}$ can be realized [101] as the isometry group of the *conformal space* $\bar{E} := E_{r+1,s+1} = E_{r,s} \perp E_{1,1}$. Therefore, the relevant algebra to work with conformal transformations of $E_{r,s}$ is its *conformal geometric algebra*, $\mathcal{G}_{r+1,s+1}$. Let us say that this hints to an important bearing on physics, particularly in the area of conformal field theory, as it can be seen in texts such as [126] and the references provided there.

It turns out to be convenient to take a basis of isotropic vectors $\boldsymbol{e}_0, \boldsymbol{e}_\infty$ of the hyperbolic plane $E_{1,1}$ such that $\boldsymbol{e}_0 \cdot \boldsymbol{e}_\infty = -1$ and then the *conformal closure* of E is the map $H : E \to \bar{E}, \boldsymbol{x} \mapsto x$ (H of Hestenes), given by the expression:

$$x = \boldsymbol{e}_0 + \boldsymbol{x} + \kappa(\boldsymbol{x})\boldsymbol{e}_\infty, \quad \kappa(\boldsymbol{x}) = \boldsymbol{x}^2/2.$$

It is immediate to check that $x \cdot y = -\frac{1}{2}(\boldsymbol{x} - \boldsymbol{y})^2$, which says that we can find the quadratic separation in E by means of the metric of $\bar{E}$. In particular, x is an isotropic vector for any $\boldsymbol{x} \in E$, and it is easy to check that the map $\boldsymbol{x} \mapsto |x\rangle$ gives a one-to-one map between E and the null cone Q in $\mathbf{P}\bar{E}$ without the point $|\boldsymbol{e}_\infty\rangle$. This gives the key to interpret the isometries of $\bar{E}$ as transformations of $E \sqcup \{|\boldsymbol{e}_\infty\rangle\}$, for an isometry of $\bar{E}$ maps Q to itself. The outcome is that in this way we get an isomorphism between the group $\mathrm{O}_{r+1,s+1}$ and the conformal group of E, $\mathrm{Conf}(E)$, and hence a 2:1 homomorphism $\mathrm{Pin}_{r+1,s+1} \to \mathrm{Conf}(E)$. The detailed proofs of all these statements follow the same pattern as those for E_3, as for example in the exposition [98, Ch. 2]. The part $\boldsymbol{e}_0 + \boldsymbol{x}$ is a realization of what in geometry books is called the projective closure of E, and it provides a means of treating projective geometry of E by means of geometric algebra: [82, 84, 102].

6.4.4 (Further Specific Applications) In geometry, recent works show that conics in a plane and quadrics in space can be encoded in $\mathcal{G}_{5,3}$ and $\mathcal{G}_{9,6}$, respectively (see [88] and [86]).

In the standard model of particle physics, a number of authors have shown the relevance of a plethora of geometric algebras for diverse purposes: see, for example [49–51, 133, 137]. Let me end this section with a quotation from this last work (p. 3): "It may also introduce many readers to the importance of Clifford algebras in particle physics, beyond their implementation in the Dirac equation."

6.5 In the Light of Lie Theory

The main aim of this section is to see how the algebraic, topological, and differentiable findings of the preceding chapters, especially those studied in Chap. 5, are placed within the framework of Lie groups' theory. Thus, it is both an entry gate to the realm of Lie theory and a source of interesting examples to accompany a deeper exploration of that realm or of related ones. In the first seven blocks, we summarize a few differential notions that were used in the preceding chapter, such as a working version of the inverse function theorem (cf. 5.4.10, p. 97).

6.5.1 (Smooth Functions, Differentials, and Directional Derivatives) Let E and F be vector spaces, U an open set of E, and $f : U \to F$ a map. We say that f is *smooth* if f has continuous partial derivatives of all orders at any point of U. This definition is independent of the basis of E with respect to which we take the partial derivatives.

The *differential* of f at $\boldsymbol{x} \in U$ is the *linear map* $d_{\boldsymbol{x}} f : E \to F$ that approximates the increment $(\Delta_{\boldsymbol{x}} f)(\boldsymbol{v}) = f(\boldsymbol{x}+\boldsymbol{v}) - f(\boldsymbol{x})$, as a function of $\boldsymbol{v}$, up to second-order terms, which formally means that

$$f(\boldsymbol{x}+\boldsymbol{v}) - f(\boldsymbol{x}) = (d_{\boldsymbol{x}} f)(\boldsymbol{v}) + o(\boldsymbol{v}), \quad \text{where } o(\boldsymbol{v})/||\boldsymbol{v}|| \to 0 \text{ when } \boldsymbol{v} \to 0.$$

For the meaning of $||\boldsymbol{v}||$, see footnote 6, p. 110. This implies that the *directional derivative* of f at $\boldsymbol{x}$ in the direction $\boldsymbol{v}$,

$$(D_{\boldsymbol{v}} f)(\boldsymbol{x}) := \tfrac{d}{dt} f(\boldsymbol{x}+t\boldsymbol{v})|_{t=0},$$

exists for any $\boldsymbol{x} \in U$ and any $\boldsymbol{v} \in E$, and that it satisfies

$$(D_{\boldsymbol{v}} f)(\boldsymbol{x}) = (d_{\boldsymbol{x}} f)(\boldsymbol{v}). \tag{6.16}$$

In fact, we have, given $\boldsymbol{x} \in U$, $\boldsymbol{v} \in E$, and t in a small interval $(-\varepsilon, \varepsilon) \subset \mathbb{R}$,

$$f(\boldsymbol{x}+t\boldsymbol{v}) - f(\boldsymbol{x}) = (d_{\boldsymbol{x}} f)(t\boldsymbol{v}) + o(t), \quad o(t)/t \to 0 \text{ for } t \to 0,$$

and it is enough to divide by t and let $t \to 0$.

6.5.2 More generally, if $Y \subseteq E$, a map $f : Y \to F$ is *smooth* if for any point $\boldsymbol{y} \in Y$ there is an open set $U_{\boldsymbol{y}} \subseteq E$ that contains $\boldsymbol{y}$ and a smooth function $\varphi_{\boldsymbol{y}} : U_{\boldsymbol{y}} \to F$ such that $f(\boldsymbol{x}) = \varphi_{\boldsymbol{y}}(\boldsymbol{x})$ for all $\boldsymbol{x} \in U_{\boldsymbol{y}} \cap Y$.

If $f : Y \to F$ is smooth and $Z = f(Y)$, we say that $f : Y \to Z$ is a *diffeomorphism* if f is bijective and $f^{-1} : Z \to Y$ is smooth.

For example, the stereographic projection $\sigma : S^{n-1} \backslash \{\boldsymbol{e}_n\} \to E_{n-1}$ considered in 6.1.2 is a diffeomorphism. Indeed, the expression for σ given in Eq. (6.1) shows that it makes sense, and is smooth, for any point not on the hyperplane $x_n = 1$, while $\sigma^{-1} : E_{n-1} \to E_n$ is also smooth and its image is $S^{n-1} \backslash \{\boldsymbol{e}_n\}$.

6.5.3 (Manifolds) A space Y is said to be a *manifold* of *dimension d* if each point $\boldsymbol{y} \in Y$ has an open neighborhood (in Y) that is diffeomorphic to an open set of E_d. The dimension d of Y is denoted by $\dim(Y)$. For example, any non-empty open set U of a vector space E_n is a manifold and $\dim U = n$, and from the preceding paragraph it follows that S^{n-1} is a manifold of dimension $n-1$ for any $n \geqslant 1$.

6.5.4 (Projective Spaces and Grassmannians) The notion of manifold in the preceding paragraph needs a broadening that liberates it from having to be a subset of some vector space (see, for instance [127, §5.1], or [47, §1.2b]). The definition of an abstract manifold is quite natural, as it is based on reflecting that it looks like an open set of a vector space around each of its points. For example, if we identify antipodal points on the sphere S^{n-1}, an operation that can be denoted $P^{n-1} := S^{n-1}/\{\pm 1\}$, we get the projective space P^{n-1}, which is a manifold in the abstract sense. Indeed, any open set of S^{n-1} that does not contain pairs of antipodal points is mapped injectively into P^{n-1}, which means that locally P^{n-1} looks like the manifold S^{n-1}. Since $P^{n-1} \simeq \mathbf{P}E_n$, we may conclude that the projective space $\mathbf{P}E_n$ is a manifold of dimension $n-1$.

More generally, the set $\mathrm{Gr}_k(E) \subset \mathbf{P}(\wedge^k E)$ introduced in Remark 2.1.15 is in fact a submanifold of dimension $(k+1)(n-k)$ of $\mathbf{P}(\wedge^k E)$. Such manifolds are called *Grassmann manifolds*, popularly *Grassmannians* [47, §17.2b].

6.5.5 (Tangent Space) If $Y \subseteq E_n$ is a manifold, and $\boldsymbol{y} \in Y$, a vector $\boldsymbol{v} \in E_n$ is said to be *tangent* to Y at $\boldsymbol{y}$ if there is a smooth function $\gamma : (-\varepsilon, \varepsilon) \to Y$, $(-\varepsilon, \varepsilon) \subset \mathbb{R}$, such that $\gamma(0) = \boldsymbol{y}$ and $\dot{\gamma}(0) = \boldsymbol{v}$. We will write $T_{\boldsymbol{y}}Y$ to denote the set of vectors tangent to Y at $\boldsymbol{y}$, and we will say that it is the *tangent space* to Y at $\boldsymbol{y}$. For example, $T_{\boldsymbol{y}}E_n = E_n$ for any point $\boldsymbol{y} \in E_n$, because if $\gamma(t) = \boldsymbol{y} + t\boldsymbol{v}$, $\boldsymbol{v} \in E_n$, then we have $\dot{\gamma}(t) = \boldsymbol{v}$ for any t. We also have seen that $T_{\mathrm{Id}}\mathrm{SO}^0_{r,s} = \mathfrak{so}_{r,s}$ (see 5.3.1), and $T_1(\mathcal{R}_{r,s}) = \mathcal{G}^2$ (see 5.4.7).

Since $\mathrm{GL}(E_n) \subset \mathrm{End}(E_n)$ is open, we have $T_{\mathrm{Id}}\mathrm{GL}(E_n) = T_{\mathrm{Id}}\mathrm{End}(E_n) = \mathrm{End}(E_n)$.

In general, as we will see in the next paragraph, $T_{\boldsymbol{a}}(X)$ is a linear subspace of E_n and $\dim T_{\boldsymbol{a}}(X) = \dim X$.

6.5.6 (Inverse Function Theorem) Let E and F be vector spaces, U a non-empty open set of E, and $f : U \to F$ a smooth function. If $\boldsymbol{u} \in U$ is such that $d_{\boldsymbol{u}}f : E \to F$ is injective, then there exists an open set $U' \subseteq U$, $\boldsymbol{u} \in U'$, such that

$f : U' \to f(U')$ is a diffeomorphism. This means that $f(U)$ is a manifold of dimension $\dim(E)$ near $f(\boldsymbol{u})$. See, for example [127, §5.3, Th. 3].

For example, since the differential of $\exp : \mathfrak{so}_{r,s} \to \mathrm{SO}_{r,s}$ is the inclusion of $\mathfrak{so}_{r,s}$ in $\mathrm{End}(E_{r,s})$, it follows that $\mathrm{SO}_{r,s}$ is a manifold near Id of dimension $\binom{r+s}{2}$ (see E.5.9, (3)). Finally, note that $\mathrm{SO}_{r,s}$ is a manifold of dimension $\binom{r+s}{2}$ near any $g \in \mathrm{SO}_{r,s}$ because the map $L_g : \mathrm{SO}_{r,s} \to \mathrm{SO}_{r,s}$, $f \mapsto gf$, is a diffeomorphism and $g = L_g(\mathrm{Id})$. Note that L_g is smooth and that its inverse is $L_{g^{-1}}$.

6.5.7 (Implicit Function Theorem) Let E and F be vector spaces, U a non-empty open set of E, and $f : U \to F$ a smooth function. Set $Z = \{z \in U \mid f(z) = 0\}$. If $z \in Z$ is such that $d_z f : E \to F$ is surjective, then there exists an open set $U' \subseteq U$, $z \in U'$, such that $Z' = Z \cap U'$ is a manifold of dimension $d = \dim(E) - \dim(F)$ and $T_z Z' = \ker(d_z f)$. See, for example [127, §5.3, Th. 4].

Although we know, via the stereographic projection, that S^{n-1} is a manifold of dimension $n-1$, it is instructive to prove it again using the implicit function theorem. Consider the function $f : E_n \to \mathbb{R}$ given by $f(\boldsymbol{x}) = \boldsymbol{x}^2$, so that $S^{n-1} = Z(f-1)$. To apply the theorem, let us find $d_{\boldsymbol{y}} f$ at a point $\boldsymbol{y} \in S^{n-1}$. By Eq. (6.16), for any vector $\boldsymbol{v} \in E_n$ we have $(d_{\boldsymbol{y}} f)(\boldsymbol{v}) = \frac{d}{dt} f(\boldsymbol{y} + t\boldsymbol{v})|_{t=0} = \frac{d}{dt}(\boldsymbol{y} + t\boldsymbol{v})^2|_{t=0} = 2\boldsymbol{y} \cdot \boldsymbol{v}$. Now for any non-zero $\boldsymbol{y}$, in particular for any $\boldsymbol{y} \in S^{n-1}$, the map $E_n \to \mathbb{R}$, $\boldsymbol{v} \mapsto 2\boldsymbol{y} \cdot \boldsymbol{v}$ is surjective. Therefore, S^{n-1} is a manifold of dimension $n-1$ around any one of its points.

As another example, consider the group $\mathrm{SL}(E) \subset \mathrm{GL}(E)$, which by definition can be represented as $Z(\det -1)$. We will see that $d_{\mathrm{Id}} \det = \mathrm{tr}$, from which it follows, since $\mathrm{tr} : \mathrm{End}(E) \to \mathbb{R}$ is surjective, that $\mathrm{SL}(E)$ is a manifold near Id of dimension $n^2 - 1$ ($n = \dim E$) and $T_{\mathrm{Id}}\mathrm{SL}(E) = \{h \in \mathrm{GL}(E) \mid \mathrm{tr}(h) = 0\}$. To prove the claim, note that for any $h \in \mathrm{End}(E)$ we have

$$(d_{\mathrm{Id}} \det)(h) = \tfrac{d}{dt} \det(\mathrm{Id} + th)|_{t=0} = \tfrac{d}{dt}(1 + \mathrm{tr}(th) + \cdots)|_{t=0} = \mathrm{tr}(h).$$

Finally, note that $\mathrm{SL}(E)$ is a manifold of dimension $n^2 - 1$ near any $g \in \mathrm{SL}(E)$ because the map $L_g : \mathrm{SL}(E) \to \mathrm{SL}(E)$, $f \mapsto gf$, is a diffeomorphism and $L_g(\mathrm{Id}) = g$.

6.5.8 (Lie Groups) We have seen that the groups GL_n, SL_n, $\mathrm{SO}_{r,s}$ are at the same time topological groups and manifolds, and that in fact the multiplication and inversion maps are smooth. In other words, they are *Lie groups*. Their dimensions are n^2, $n^2 - 1$, and $\binom{r+s}{2}$, respectively.

The spinorial groups are also Lie groups. To see this, let us look first at the case $\mathcal{G}^\times = \mathcal{G}^\times_{r,s}$, the group of invertible elements of $\mathcal{G} = \mathcal{G}_{r,s}$. Consider the exponential map $\exp : \mathcal{G} \to \mathcal{G}^\times$, $\exp(x) = e^x$. This map is smooth (it is actually analytic) and $d_0 \exp : \mathcal{G} \to \mathcal{G}$ is the identity. Indeed, for any multivector x,

$$(d_0 \exp)(x) = \tfrac{d}{dt} e^{tx}|_{t=0} = x.$$

By the inverse function theorem, there is an open set U in $\mathcal{G}$ containing 0 such that $\exp : U \to \exp(U)$ is a diffeomorphism. In particular, $\exp(U) \subseteq \mathcal{G}^{\times}$ is open in $\mathcal{G}$, and this implies that $\mathcal{G}^{\times}$ is open in $\mathcal{G}$, for if $u \in \mathcal{G}^{\times}$, then $u\exp(U) \subseteq \mathcal{G}^{\times}$ is open in $\mathcal{G}$. Since the product is smooth, and $(ue^x)^{-1} = e^{-x}u^{-1}$ for any $x \in U$, we can conclude that $\mathcal{G}^{\times}$ is a Lie group of dimension $\dim \mathcal{G} = 2^n$. Note that this also completes the discussion in 6.1.6 to show that $\mathcal{G}^{\times}$ is a topological group.

In the case of the rotor group $\mathcal{R}_{r,s}$, the proof that it is a Lie group of dimension $\binom{r+s}{2}$ is similar to the argument for $\mathcal{G}^{\times}$, as we know that the differential of exponential map $\exp : \mathcal{G}^2 \to \mathcal{R}_{r,s}$ is the identity (see 5.4.10). From this, together with 4.3.3 and 4.3.4, it is clear that the groups $\mathrm{Spin}_{r,s}$ and $\mathrm{Pin}_{r,s}$ are also Lie groups of dimension $\binom{r+s}{2}$. Finally, the versor group $\mathcal{V}_{r,s}$ is also a Lie group of dimension $1 + \binom{r+s}{2}$. This can be proved using 4.2.2(4) (there is one extra degree of freedom stemming from the kernel $\mathbb{R}^{\times}$), but it is more enlightening to consider the exponential map:

$$\exp : \mathbb{R} + \mathcal{G}^2 \to \mathcal{G}.$$

Since $e^{\lambda+b} = e^{\lambda}e^{b}$ is a versor ($\lambda \in \mathbb{R}$, $b \in \mathcal{G}^2$), we actually have an exponential map:

$$\exp : \mathbb{R} + \mathcal{G}^2 \to \mathcal{V}_{r,s}$$

and we can adapt the arguments so far to see that it induces a diffeomorphism of an open set U of $\mathbb{R} + \mathcal{G}^2$ containing 0 with an open set U' of $\mathcal{V}_{r,s}$ containing 1 and then carry on as in the other cases.

6.5.9 (The Lie Algebra) Let G be any of the groups considered in the two preceding paragraphs, and write $\mathfrak{lie}(G)$ to denote its tangent space at the identity element of G. More specifically, we have

$$\mathfrak{lie}(\mathrm{GL}(E)) = \mathrm{End}(E) \tag{6.17}$$

$$\mathfrak{lie}(\mathrm{SL}(E)) = \mathrm{End}_0(E) \text{ (the traceless endomorphisms of} E) \tag{6.18}$$

$$\mathfrak{lie}(\mathrm{O}_{r,s}) = \mathfrak{lie}(\mathrm{SO}_{r,s}) = \mathfrak{lie}(\mathrm{SO}^0_{r,s}) = \mathfrak{so}_{r,s} \tag{6.19}$$

$$\mathfrak{lie}(\mathcal{G}^{\times}_{r,s}) = \mathcal{G}_{r,s} \tag{6.20}$$

$$\mathfrak{lie}(\mathrm{Pin}_{r,s}) = \mathfrak{lie}(\mathrm{Spin}_{r,s}) = \mathfrak{lie}(\mathcal{R}_{r,s}) = \mathcal{G}^2_{r,s} \tag{6.21}$$

$$\mathfrak{lie}(\mathcal{V}_{r,s}) = \mathbb{R} + \mathcal{G}^2_{r,s} \tag{6.22}$$

In all cases, $\mathfrak{lie}(G)$ is closed under the commutator bracket, and hence it is a Lie algebra. The only cases in which this may not be obvious can be justified as follows. Since the commutator of any two endomorphisms is traceless, a fortiori we have that $\mathrm{End}_0(E)$ is closed under the commutator bracket. The case of $\mathfrak{so}_{r,s} = \mathfrak{so}(E_{r,s})$ has been established in 5.3.4. Finally, 5.4.3 tells us that $\mathcal{G}^2_{r,s}$ is also closed under the commutator bracket. Note that this implies that $\mathbb{R} + \mathcal{G}^2_{r,s}$ has the same property: $[\lambda + b, \lambda' + b'] = [b, b']$ for $\lambda, \lambda' \in \mathbb{R}$ and $b, b' \in \mathcal{G}^2_{r,s}$.

In all cases, we have an exponential map:

$$\exp : \mathfrak{lie}(G) \to G, \quad x \mapsto e^x,$$

which provides the following interpretation of the commutator bracket:

$$[x, y] = \tfrac{d}{dt}(e^{tx} y e^{-tx})|_{t=0}.$$

Here, the main point is that the expression $e^{tx} y e^{-tx}$ makes sense in all cases. Take, for example, $G = \mathcal{R}_{r,s}$. Then, $x, y \in \mathcal{G}^2_{r,s}$, $e^{tx} \in \mathcal{R}_{r,s}$, and we know (see 5.4.6) that the map $\mathcal{G}_{r,s} \to \mathcal{G}_{r,s}$, $y \mapsto e^{tx} y e^{-tx}$, is a grade preserving algebra automorphism. The case $x, y \in \mathfrak{so}_{r,s}$ is also illustrative: $e^{tx} \in \mathrm{SO}_{r,s}$ and it is immediate to check that $e^{tx} y e^{-tx}$ is skew-symmetric if y is skew-symmetric.

6.5.10 (When Is the Exponential onto?) In 5.4.18, we have seen that the exponential map $\exp : \mathfrak{so}_n \to \mathrm{SO}_n$ is onto and in 5.4.19 that for any $R \in \mathcal{R}_n$, either R of $-R$ has the form e^b, $b \in \mathcal{G}^2_n$. This implies that if $R \in \mathcal{R}_n$ is not exponential, then $R^2 = (-R)^2$ is exponential.

For general signatures (r, s), we recall the following result (see [36, 37, 112]): The exponential of $\mathrm{SO}^0_{r,s}$ is surjective if and only if $\min(r, s) = 0, 1$, and if $\min(r, s) \geqslant 2$, then each element of $\mathrm{SO}^0_{r,s}$ is either exponential or its square is exponential. The first part of this statement is closely related to the following theorem of Riesz: If (r, s) has one of the forms $(n, 0)$, $(0, n)$, $(1, n-1)$, or $(n-1, 1)$ and $L \in \mathrm{SO}^0_{r,s}$, there exists a bivector $b \in \mathcal{G}^2_{r,s}$ such that

$$L\boldsymbol{x} = e^b \boldsymbol{x} e^{-b}.$$

Moreover, this statement is not true for any other signature [122, § 4.12].

6.5.11 (Universal Coverings) For the Euclidean spaces of dimension $n \geqslant 2$, we have the 2:1 homomorphism $\mathrm{Spin}_n = \mathcal{R}_n \to \mathrm{SO}_n$, and we know that $\mathrm{Spin}_n = \mathcal{R}_n$ is simply connected for $n \geqslant 3$ (see 6.1.7 (8)). A similar situation occurs for the signatures $(1, n)$: For $n \geqslant 2$, we have the 2:1 homomorphism $\mathcal{R}_{1,n} \to \mathrm{SO}^0_{1,n}$, and by 6.1.7 (14) the group $\mathcal{R}_{1,n}$ is simply connected for $n \geqslant 3$.

These results are special cases of *universal covering* groups. Given a connected Lie group G, there exists a simply connected Lie group $\tilde{G}$ endowed with an onto homomorphism $\tilde{G} \to G$ such that the induced map $\mathfrak{lie}(\tilde{G}) \to \mathfrak{lie}(G)$ is an isomorphism. These data are unique up to a canonical isomorphism (see [62, 2nd ed., Prop. 6.13], or, more generally, [117, § 51]). Thus, we may write

$$\widetilde{\mathrm{SO}_n} = \mathcal{R}_n \text{ and } \widetilde{\mathrm{SO}^0_{1,n}} = \mathcal{R}_{1,n} \text{ for all } n \geqslant 3. \tag{6.23}$$

For $n = 2$, the situation is quite different. In the Euclidean case, we have the onto homomorphism $\mathbb{R} \to \mathrm{U}_1 = \mathrm{SO}_2$, $\theta \mapsto e^{-\theta \boldsymbol{i}}$ (see 5.1.5). Hence, $\widetilde{\mathrm{SO}_2} = \mathbb{R}$. In the Lorentzian case, the exceptions to (6.23) are $\mathcal{R}_{1,1}$ and $\mathcal{R}_{1,2}$. As seen in 5.1.6, the group $\mathcal{R}_{1,1}$ has two connected components, and the connected component of 1 is isomorphic to $\mathbb{R}$. Hence, we have $\widetilde{\mathrm{SO}^0_{1,1}} = \mathbb{R}$. The analysis of $\mathcal{R}_{1,2}$ has been carried out in 5.1.11, where we have seen that it is isomorphic to SL_2. Moreover, we have seen there that it is homeomorphic to $\mathrm{U}_1 \times \Delta$, $\Delta \subset \mathbb{C}$ the open disc of radius 1, and hence

$$\pi_1(\mathcal{R}_{1,2}) = \pi_1(\mathrm{SL}_2) \simeq \pi_1(\mathrm{U}_1) \simeq \mathbb{Z}.$$

For a different approach to this result, based on the (unique) polar decomposition of a matrix $A \in \mathrm{SL}_2$ in the form $A = Re^S$, with $R \in \mathrm{SO}_2$ and $S \in \mathcal{S}_0(2)$, where $\mathcal{S}_0(2)$ is the vector subspace of $\mathbb{R}(2)$ formed by the traceless symmetric matrices, see [62, 2nd ed., Prop. 2.19].

The study of SL_2 is a surprisingly rich area on account of its many deep connections with other fields. Among the multitude of references, let us just mention the introductory [95], the references therein (in particular [94], where the algebras $\mathcal{G}_{0,2}$, $\mathcal{G}_{1,1}$, and $\mathcal{G}_{0,1,0}$ play a prominent role), and texts such as [25, particularly § II.2 and § II.16] and [97].

Equation (6.23) for $n = 3$ tells us that $\widetilde{\mathrm{SO}_3} = \mathcal{R}_3$ and $\widetilde{\mathrm{SO}^0_{1,3}} = \mathcal{R}_{1,3}$. The group $\mathcal{R}_3$ has been studied in 5.1.7, and we have found that it is the group of unit quaternions, which is also denoted SU_2. Let us end this block by showing that

$$\mathrm{Spin}^0_{1,3} = \mathcal{R}_{1,3} \simeq \mathrm{SL}_2(\mathbb{C}).$$

This follows by combining the isomorphism $\mathcal{G}_3 \simeq \mathcal{G}^+_{1,3}$ obtained using E.3.8, p. 60, and the isomorphism $\mathcal{G}_3 \simeq \mathbb{C}(2)$ established in E.3.7, p. 60. The first isomorphism depends on choosing a positive unit vector $\boldsymbol{e}_0 \in E_{1,3}$ and taking the associated split:

$$E_{1,3} = \langle \boldsymbol{e}_0 \rangle \perp E_{\bar{3}},$$

where $E_{\bar{3}} = \boldsymbol{e}_0^{\perp}$. Let E_3 be the same space $E_{\bar{3}}$ but with the opposite metric. To handle this gracefully, it is enough to write $\bar{\boldsymbol{x}}$ to denote a vector $\boldsymbol{x} \in E_3$ when considered as a vector in $E_{\bar{3}}$, so that $\bar{\boldsymbol{x}}^2 = -\boldsymbol{x}^2$. With this convention, the first isomorphism is the unique algebra homomorphism that extends the linear map $E_3 \to \mathcal{G}^+_{1,3}$ such that $\boldsymbol{x} \mapsto \bar{\boldsymbol{x}}\boldsymbol{e}_0$. This works because $(\bar{\boldsymbol{x}}\boldsymbol{e}_0)^2 = \bar{\boldsymbol{x}}\boldsymbol{e}_0\bar{\boldsymbol{x}}\boldsymbol{e}_0 = -\bar{\boldsymbol{x}}^2 = \boldsymbol{x}^2$. Now, note that $\widetilde{\bar{\boldsymbol{x}}\boldsymbol{e}_0} = -\bar{\boldsymbol{x}}\boldsymbol{e}_0$, so that the reverse involution in $\mathcal{G}^+_{1,3}$ becomes the Clifford conjugation in $\mathcal{G}_3$ (the unique anti-automorphism that reverses the sign of vectors). It follows that the reverse of

$$R = \xi_0 + \xi_1\boldsymbol{e}_1 + \xi_2\boldsymbol{e}_2 + \xi_3\boldsymbol{e}_3 \in \mathcal{G}_3$$

(see E.3.8 for the notations) is

$$\tilde{R} = \xi_0 - \xi_1 \boldsymbol{e}_1 - \xi_2 \boldsymbol{e}_2 - \xi_3 \boldsymbol{e}_3,$$

because $\boldsymbol{i} = \boldsymbol{e}_1\boldsymbol{e}_2\boldsymbol{e}_3$ is central and (Clifford) self-conjugate. Therefore, the condition $R\tilde{R} = 1$ for R to be a rotor is that $\xi_0^2 - \xi_1^2 - \xi_2^2 - \xi_3^2 = 1$. But, this agrees with the condition $\det(L) = 1$, where L is the matrix associated to R, namely

$$L = \begin{pmatrix} \xi_0 + \xi_3 & \xi_1 - i\xi_2 \\ \xi_1 + i\xi_2 & \xi_0 - \xi_3 \end{pmatrix}.$$

Since L^{-1} is the representation of $\tilde{R}$, as it is easy to check, and the action of $\underline{R} \in \mathrm{SO}^0_{1,3}$ on a vector $x \in E_{1,3}$ is given by $\underline{R}(x) = Rx\tilde{R}$, it turns out that in terms of the matrix representation:

$$X = \begin{pmatrix} x_0 + x_3 & x_1 - ix_2 \\ x_1 + ix_2 & x_0 - x_3 \end{pmatrix}.$$

of $x = x_0\boldsymbol{e}_0 + x_1\boldsymbol{e}_1 + x_2\boldsymbol{e}_2 + x_3\boldsymbol{e}_3$, that action is expressed by LXL^{-1}. Notice also that $q(x) = \det(X)$, where q in the metric of $E_{1,3}$, and the fact that $\det(LXL^{-1}) = \det(X)$ just confirms that $\underline{R}$ is an isometry.

Coda

To end this brief, let us point out a few references that reflect on the nature of mathematics and its relations to mathematical physics or to applications of a more general character.

6.5.12 (What Is Mathematics?) Mac Lane's 1986 book [107] is presented as "a description of the form and function of Mathematics, as a background for the Philosophy of Mathematics." It offers an accessible account of what mathematics is as a whole and what are the forces driving its unfolding. It ends with this judgment: "Mathematics aims to understand, to manipulate, to develop, and to apply those aspects of the universe which are formal. This view, as expounded in this book, might be called *formal functionalism*."

The formal corner of the universe addressed in our brief, labeled "real spinorial groups" in the title, has tried to show the deep role played by the rotor groups $\mathcal{R}_{r,s}$, hence also by the geometric algebra $\mathcal{G}_{r,s}$, in the study of the special orthochronous rotation groups $\mathrm{SO}^0_{r,s}$. Their algebraic, topological, and differentiable aspects, as well as its significance for geometry, physics, and diverse applications, surely constitute a compelling illustration of formal functionalism.

6.5.13 (A Road to Reality?) Although Mac Lane's book also discusses the applicability of mathematics to Mechanics, a much thorough discussion is assembled in Penrose's book [114]. For the present context, its last chapter, "34 Where lies the road to reality?," may be enlightening in what concerns the relation of mathematics to present day theoretical physics.

For example, the following quotation is extracted from §34.5, p. 1026: "The more deeply we probe the fundamentals of physical behavior, the more we find that it is very precisely controlled by mathematics. Moreover, the mathematics that we find is not just of a direct calculational nature; it is of a profoundly sophisticated character, where there is subtlety and beauty of a kind that is not to be seen in the mathematics that is relevant to physics at a less fundamental level. In accordance with this, progress towards a deeper physical understanding, if it is not able to be guided in detail by experiment, must rely more and more heavily on an ability to appreciate the physical relevance and depth of the mathematics, and to "sniff out" the appropriate ideas by use of a profoundly sensitive esthetic mathematical appreciation."

6.5.14 (Other Readings) As for texts offering mathematical background of the kind used in this book, the reader may like to consider Porteus's systematic book [118], "an effective course [...] which contains equal parts of linear algebra and analysis, with some of the more interesting geometrical examples included as illustrations," or Stillwell's [132], which "gives a bird's eye view of undergraduate mathematics and a glimpse of wider horizons," with the added appeal provided by the chapter previews, historical perspectives, and biographical information. For Topology, in particular for the fundamental group, the book [127] is an excellent choice.

Finally, a list of some excellent texts, not cited already and ordered by the year of the first publication, that deal, in one way or another, with Lie groups, Lie algebras, and representation theory, often with applications to Physics: [4, 14, 16, 18, 20, 25, 43, 45, 48, 52, 53, 55–57, 91–93, 108, 110, 121, 123, 125, 131, 136, 139].

References

1. R. Abłamowicz (ed.), *Clifford Algebras: Applications to Mathematics, Physics, and Engineering*. Progress in Mathematical Physics, vol. 34 (Birkhäuser, Boston, 2004)
2. W. Achtner, From religion to dialectics and mathematics. Stud. Log. Grammar Rhetor. **44**(1), 111–131 (2016)
3. J.F. Adams, *Lectures on Lie Groups* (Benjamin, Reading, 1969)
4. P. Angles, *Conformal Groups in Geometry and Spin Structures*. Progress in Mathematical Physics, vol. 50 (Birkhäuser, Boston, 2008), xvii+283 pp.
5. M. Arrayás, D. Bouwmeester, J.L. Trueba, Knots in electromagnetism. Phys. Rep. **667**, 1–61 (2017)
6. E. Artin, *Geometric Algebra*. Tracts in Pure and Applied Mathematics, vol. 3 (Interscience Publishers, New York, 1957)
7. M. Atiyah, R. Bott, A. Shapiro, Clifford modules. Topology **3**(Suppl. 1), 3–38 (1964)
8. W.E. Baylis, *Electrodynamics. A Modern Geometric Approach*. Progress in Mathematical Physics (Birkhäuser, Boston, 1999)
9. E. Bayro-Corrochano, *Geometric Computing for Perception Action Systems: Concepts, Algorithms, and Scientific Applications* (Springer, New York, 2001)
10. E. Bayro-Corrochano, *Handbook of Geometric Computing: Applications in Pattern Recognition, Computer Vision, Neuralcomputing, and Robotics* (Springer, Berlin, 2005), xv+779 pp.
11. E. Bayro-Corrochano, *Geometric Computing: For Wavelet Transforms, Robot Vision, Learning, Control and Action* (Springer, London, 2010)
12. E. Bayro-Corrochano, G. Scheuermann (eds.), *Geometric Algebra Computing: In Engineering and Computer Science* (Springer, London, 2010)
13. E. Bayro-Corrochano, G. Sobczyk, *Geometric Algebra with Applications in Science and Engineering* (Birkhäuser, Boston, 2001)
14. G. Bellamy, *Lie Groups, Lie Algebras, and Their Representations*, 2016. Lecture notes available at https://www.maths.gla.ac.uk/~gbellamy/lie.pdf
15. I.M. Benn, R.W. Tucker, *An Introduction to Spinors and Geometry with Applications in Physics* (Adam Hilger, Bristol, 1987), x+358 pp.
16. J.P. Bourguignon, O. Hijazi, J.-L. Milhorat, A. Moroianu, *A Spinorial Approach to Riemannian and Conformal Geometry*. Monographs in Mathematics (European Mathematical Society, Zürich, 2015)
17. R. Brauer, H. Weyl, Spinors in n dimensions. Am. J. Math. **57**, 230–254 (1935)
18. T. Bröcker, T. Dieck, *Representations of Compact Lie Groups*. Graduate Texts in Mathematics, vol. 98 (Springer, New York, 1985)

S. Xambó-Descamps, *Real Spinorial Groups*, SpringerBriefs in Mathematics,
https://doi.org/10.1007/978-3-030-00404-0

19. P. Budinich, A. Trautman, *The Spinorial Chessboard* (Springer, Berlin, 1988)
20. D. Bump, *Lie Groups*. Graduate Texts in Mathematics, vol. 225 (Springer, Berlin, 2004)
21. J. Cameron, J. Lasenby, Oriented conformal geometric algebra. Adv. Appl. Clifford Algebr. **18**(3–4), 523–538 (2008)
22. J.F. Cariñena, A. Ibort, G. Marmo, G. Morandi, *Geometry from Dynamics, Classical and Quantum* (Springer, Berlin, 2015)
23. É. Cartan, Les groupes projectifs qui ne laissent invariante aucune multiplicité plane. Bull. Soc. Math. Fr. **41**, 53–96 (1913)
24. É. Cartan, *Leçons sur la théorie des spineurs* (2 volumes). Actualités Scientifiques et Industrielles, vol. 643 (Hermann, Paris, 1937). Translated into English in 1966 (The theory of spinors, Hermann), and republished by Dover in 1981
25. R. Carter, G. Segal, I. Macdonald, *Lectures on Lie Groups and Lie Algebras*. LMS Student Texts, vol. 32 (Cambridge University Press, Cambridge, 1995). The book has three parts: Lie algebras and root systems (Carter, 1–44); Lie groups (Segal, 45–132); and Linear algebraic groups (Macdonald, 133–188)
26. D. Castelvecchi, The shape of things to come. Nature **547**, 272–274 (2017)
27. C. Chevalley, *The Algebraic Theory of Spinors* (Columbia University Press, New York, 1954), viii+131 pp.
28. C. Chevalley, *The Construction and Study of Certain Important Algebras* (The Mathematical Society of Japan, Tokyo, 1955), vi+64 pp.
29. C. Chevalley, *The Algebraic Theory of Spinors and Clifford Algebras*. Collected Works, vol. 2 (Springer, Berlin, 1997). Includes [28] and [27], a Preface by C. Chevalley and P. Cartier, the review of [27] by J. Dieudonné, and a Postface by J.-P. Bourguignon with the title "Spinors in 1995"
30. W.K. Clifford, Applications of Grassmann's extensive algebra. Am. J. Math. **1**, 350–358 (1878)
31. P. Colapinto, Spatial computing with conformal geometric algebra. Master's thesis, University of California Santa Barbara, 2011. http://wolftype.com/versor/colapinto_masters_final_02.pdf
32. P. Colapinto, Articulating space: geometric algebra for parametric design – symmetry, kinematics, and curvature. PhD thesis, Media Arts and Technology Program, University of California Santa Barbara, 2016
33. A. Crumeyrolle, *Orthogonal and Symplectic Clifford Algebras – Spinor Structures* (Kluwer Academic Publishers, Dordrecht, 1990)
34. J. Dieudonné, *Sur le groupes classiques*. Actualités scientifiques et industrielles, vol. 1040 (Hermann, Paris, 1973)
35. P.A.M. Dirac, The quantum theory of the electron, I, II. Proc. R. Soc. Lond. **118**, A117:610–624 and A118:351–361 (1928)
36. D.Ž. Đogović, K.H. Hofmann, The surjectivity question for the exponential function of real Lie groups: a status report. J. Lie Theory **7**, 171–199 (1997)
37. D.Z. Doković, The interior and the exterior of the image of the exponential map in classical Lie groups. J. Algebra **112**, 90–109 (1988)
38. C. Doran, A. Lasenby, *Geometric Algebra for Physicists* (Cambridge University Press, Cambridge, 2003)
39. L. Dorst, Structure preserving representation of Euclidean motions through Conformal Geometric Algebra, in *Guide to Geometric Algebra in Practice*, ed. by L. Dorst, J. Lasenby (Springer, Berlin, 2011)
40. L. Dorst, J. Lasenby (eds.), *Guide to Geometric Algebra in Practice* (Springer, Berlin, 2011)
41. L. Dorst, C.J.L. Doran, J. Lasenby (eds.), *Applications of Geometric Algebra in Computer Science and Engineering* (Springer, Berlin, 2002)
42. L. Dorst, D. Fontijne, S. Mann, *Geometric Algebra for Computer Science: An Object-Oriented Approach to Geometry* (Morgan Kaufmann Publishers Inc., Amsterdam, 2007)
43. J.J. Duistermaat, J.A.C. Kolk, *Lie Groups* (Springer, Berlin, 2012)

44. J. Figueroa-O'Farrill, Spin geometry. http://empg.maths.ed.ac.uk/Activities/Spin/, versión 4/5/2010
45. G.B. Folland, *Quantum Field Theory. A Tourist Guide for Mathematicians*. Mathematical Surveys and Monographs, vol. 149 (American Mathematical Society, Providence, 2008)
46. M. Forster, Friedrich Daniel Ernst Schleiermacher, in *The Stanford Encyclopedia of Philosophy*, ed. by E.N. Zalta, 2015 edition. https://plato.stanford.edu/archives/sum2015/entries/schleiermacher
47. T. Frankel, *The Geometry of Physics. An Introduction* (Cambridge University Press, Cambridge, 2004)
48. W. Fulton, J. Harris, *Representation Theory. A First Course*. Graduate Texts in Mathematics (Springer, Berlin, 1991)
49. C. Furey, Charge quantization from a number operator. Phys. Lett. B **742**, 195–199 (2015)
50. C. Furey, Standard model physics from an algebra? Ph.D. thesis, University of Waterloo, 2015. www.repository.com.ac.uk/handle/1810/254719
51. C. Furey, A demonstration that electroweak theory can violate parity automatically (leptonic case). Int. J. Mod. Phys. A **33**(4), 1830005 (10 pp.) (2018)
52. J.H. Gallier, *Geometric Methods and Applications, For Computer Science and Engineering*. Texts in Applied Mathematics, vol. 38, 2nd edn. (Springer, Berlin, 2011)
53. J.H. Gallier, J. Quaintance, *Notes on Differential Geometry of Lie Groups*. Texts in Applied Mathematics, vol. 38 (Springer, Berlin, 2016)
54. D.J.H. Garling, *Clifford Algebras: An Introduction*. LMS Student Texts, vol. 78 (Cambridge University Press, Cambridge, 2011)
55. H. Georgi, *Lie Algebras in Particle Physics* (Westview/Perseus Books, Reading, 1999)
56. R. Goodman, N.R. Wallach, *Symmetry, Representations, and Invariants*. Graduate Texts in Mathematics, vol. 65 (Springer, Dordrecht, 2009)
57. M. Gourdin, *Basics of Lie Groups* (Frontières, Gif-sur-Yvette, 1982)
58. H.G. Grassmann, *Die lineale Ausdehnungslehre, ein neuer Zweig der Mathematik* (Otto Wiegand, Leipzig, 1844)
59. H.G. Grassmann, *Die Ausdehnungslehre. Vollständig und in strenger Form* (Adolf Enslin, Berlin, 1862)
60. H.G. Grassmann, *Extension Theory* (American Mathematical Society, Providence, 2000). Translated from the German version *Die Ausdehnungslehre von 1862* by Lloys C. Kannenberg
61. L.C. Grove, *Classical Groups and Geometric Algebra*. Graduate Studies in Mathematics, vol. 29 (American Mathematical Society, Providence, 2002)
62. B. Hall, *Lie Groups, Lie Algebras, and Representations: An Elementary Introduction*. Graduate Texts in Mathematics, vol. 222 (Springer, Cham, 2003), xiv+351 pp. Second edition 2015, xiv+451 pp.
63. W.R. Hamilton, *Lectures on Quaternions* (Hodges and Smith, Dublin, 1853). http://ebooks.library.cornell.edu/m/math/index.php
64. T.F. Havel, Distance geometry: theory, algorithms, and chemical applications. Encycl. Comput. Chem. **120**, 723–742 (1998)
65. D. Hestenes, *Space-Time Algebra* (Gordon & Breach, New York, 1966). 2nd edition: Birkhäuser 2015, with a Foreword by A. Lasenby and new "Preface after fifty years" by the author
66. D. Hestenes, A unified language for mathematics and physics, in *Clifford Algebras and Their Applications in Mathematical Physics*, ed. by J.S.R. Chisholm, A.K. Commons (Reidel, Dordrecht, 1986), pp. 1–23
67. D. Hestenes, Grassmann's vision, in *Hermann Gunther Grasmann (1809–1877): Visionary Mathematician, Scientist and Neohumanist Scholar*, ed. by G. Schubring (Kluwer, Boston, 1996), pp. 191–201
68. D. Hestenes, Real Dirac theory, 1996, in *The Theory of the Electron*, ed. by J. Keller, Z. Oziewicz (UNAM, México, 1996), pp. 1–50
69. D. Hestenes, *New Foundations for Classical Mechanics*. Fundamental Theories of Physics, vol. 99, 2nd edn. (Kluwer Academic Publishers, Dordrecht, 1999). 1st edition published 1990

70. D. Hestenes, Old wine in new bottles: a new algebraic framework for computational geometry, in *Advances in Geometric Algebra with Applications in Science and Engineering*, ed. by E. Bayro-Corrochano, G. Sobczyk (Birkhäuser, Boston, 2001), pp. 1–14
71. D. Hestenes, Point groups and space groups in geometric algebra, in *Applications of Geometric Algebra in Computer Science and Engineering*, ed. by L. Dorst, C. Doran, J. Lasenby (Birkhäuser, Boston, 2002), pp. 3–34
72. D. Hestenes, Mysteries and insights of Dirac theory. Ann. Fond. Louis de Broglie **28**(3), 390–408 (2003)
73. D. Hestenes, Oersted Medal Lecture 2002: reforming the mathematical language of physics. Am. J. Phys. **71**(2), 104–121 (2003)
74. D. Hestenes, New tools for computational geometry and rejuvenation of screw theory, in *Geometric Algebra Computing in Engineering and Computer Science*, ed. by E. Bayro-Corrochano, G. Scheuermann (Springer, London, 2010), pp. 3–34
75. D. Hestenes, Grassmann's legacy, in *From Past to Future: Grassmann's Work in Context*, ed. by H.-J. Petsche, A. Lewis, J. Liesen, S. Russ (Birkhäuser, Basel, 2011), pp. 243–260
76. D. Hestenes, The shape of differential geometry in geometric calculus, in *Guide to Geometric Algebra in Practice*, ed. by L. Dorst, J. Lasenby (Springer, London, 2011), pp. 393–410
77. D. Hestenes, The genesis of geometric algebra: a personal perspective. Adv. Appl. Clifford Algebr. **27**(1), 351–379 (2017)
78. D. Hestenes, Deconstructing the electron clock, 2018. Preprint received on July 16, 2018. Can be accessed at http://www.ime.unicamp.br/~agacse2018/abstracts/InvitedSpeakers/Hestenes-Maxwell-Dirac.pdf
79. D. Hestenes, Quantum Mechanics of the electron particle-clock, 2018. Preprint received on July 16, 2018. Can be accessed at http://www.ime.unicamp.br/~agacse2018/abstracts/InvitedSpeakers/Hestenes-ElectronClock.pdf
80. D. Hestenes, J. Holt, The crystallographic space groups in geometric algebra. J. Math. Phys. **48**, 023514 (2007)
81. D. Hestenes, G. Sobczyk, *Clifford Algebra to Geometric Calculus* (Reidel, Dordrecht, 1984)
82. D. Hestenes, R. Ziegler, Projective geometry with Clifford algebra. Acta Appl. Math. **23**, 25–63 (1991)
83. D. Hestenes, H. Li, A. Rockwood, Spherical conformal geometry with geometric algebra, in *Geometric Computing with Clifford Algebras* (Springer, Berlin, 2001), pp. 61–75
84. D. Hestenes, H. Li, A. Rockwood, New algebraic tools for classical geometry, in *Geometric Computing with Clifford Algebras* (Springer, Berlin, 2001), pp. 3–26
85. D. Hildenbrand, *Foundations of Geometric Algebra Computing* (Springer, Berlin, 2012)
86. E. Hitzer, Three-dimensional quadrics in hybrid conformal geometric algebras of higher dimensions, in *Early Proceedings of AGACSE 2018* (2018)
87. J. Hladic, *Spinors in Physics*. Graduate Texts in Contemporary Physics (Springer, New York, 1999). Translated by J. Michael Cole from the French edition "Les spineurs en physique" published in 1996 by Masson, Paris
88. J. Hrdina, A. Návrat, P. Vasik, Geometric algebra of conics (2018). Preprint
89. S. Huang, Y.Y. Qiao, G.C. Wen, *Real and Complex Clifford Analysis*. Advances in Complex Analysis and Its Applications (Springer, Berlin, 2006)
90. B. Jancewicz, *Multivectors and Clifford Algebra in Electrodynamics* (World Scientific, Singapore, 1988), xiv+316 pp.
91. N. Jeevanjee, *An Introduction to Tensors and Group Theory for Physicists*, 2nd edn. (Birkhäuser, Cham, 2014)
92. L. Ji, A. Papadopoulos (eds.), *Sophus Lie and Felix Klein: The Erlangen Program and Its Impact in Mathematics and in Physics*, vol. 23 (European Mathematical Society, Zürich, 2015). xviii+330 pp.
93. A. Kirillov Jr., *An Introduction to Lie Groups and Lie Algebras*. Cambridge Studies in Advanced Mathematics, vol. 113 (Cambridge University Press, Cambridge, 2008)
94. V.V. Kisil, Elliptic, parabolic and hyperbolic analytic function theory-1: geometry of invariants, 2006. https://arxiv.org/pdf/math/0512416v4.pdf

95. V.V. Kisil, Starting with the Group $SL_2(\mathbb{R})$. Not. AMS **54**(11), 1458–1465 (2007)
96. Y. Kuroe, T. Nitta, E. Hitzer, Applications of Clifford's geometric algebra. SICE J. Control Meas. Syst. Integr. **4**(1), 1–10 (2011)
97. S. Lang, $SL_2(\mathbf{R})$. Graduate Texts in Mathematics (Springer, Berlin, 1985)
98. C. Lavor, S. Xambó-Descamps, I. Zaplana, *A Geometric Algebra Invitation to Space-Time Physics, Robotics and Molecular Geometry*. SBMA/Springerbrief (Springer, Cham, 2018)
99. H.B. Lawson, M.-L. Michelsohn, *Spin Geometry* (Princeton University Press, Princeton, 1989)
100. H. Li, Symbolic geometric reasoning with advanced invariant algebras, in *International Conference on Mathematical Aspects of Computer and Information Sciences* (Springer, Cham, 2015), pp. 35–49
101. H. Li, D. Hestenes, A. Rockwood, A universal model for conformal geometries of Euclidean, spherical and double-hyperbolic spaces, in *Geometric Computing with Clifford Algebras* (Springer, Berlin, 2001), pp. 77–104
102. H. Li, D. Hestenes, A. Rockwood, Generalized homogeneous coordinates for computational geometry, in *Geometric Computing with Clifford Algebras* (Springer, Berlin, 2001), pp. 27–59
103. R. Lipschitz, Principes d'un calcul algébrique que contient comme espèces particulières le calcul des quantités imaginaires et des quaternions. C. R. Acad. Sci. Paris **xli** (1880). Reproduced in the Bull. Sci. Math. (2) **11**, 115–120 (1887)
104. R. Lipschitz, *Untersuchungen über die Summen von Quadraten* (M. Cohen and Sohn, Pittsburgh, 1886). The first chapter, pp. 5–57. Translated into French by J. Molk: *Recherches sur la transformation, par des substitutions réelles, d'un somme de deux ou trois carrés en elle-mêmme*, J. Math. Pures Appl. **2**, 163–183 (1886)
105. P. Lounesto, *Clifford Algebras and Spinors. LMS Lecture Notes Series*, vol. 286, 2nd edn. (Cambridge University Press, Cambridge, 2001)
106. D. Lundholm, L. Svensson, Clifford algebra, geometric algebra, and applications (2009), Updated 2016. http://arxiv.org/pdf/0907.5356.pdf
107. S. MacLane, *Mathematics Form and Function* (Springer, New York, 1986)
108. E. Meinrenken, A.M. Cohen, *Clifford Algebras and Lie Theory* (Springer, Berlin, 2013)
109. H. Minkowski, Space and time, in *The Principle of Relativity* (Dover, New York, 1952). Translation of the communication«Raum und Zeit» presented by the author to the 80th Convention of German Scientists and Doctors (Köln, 21 September 1908)
110. R. Mneimné, F. Testard, *Introduction à la théorie des groupes classiques*. Méthodes (Hermann, Paris, 1986)
111. J.A. Navarro, *Notes for a Degree in Mathematics*. Algebra and Geometry (2017). http://matematicas.unex.es/~navarro/degree.pdf. Based on Lectures of J. Sancho
112. M. Nishikawa, On the exponential map of the group $O(p,q)_0$. Mem. Fac. Sci. Kyushu Univ. **37**(1), 63–69 (1983)
113. W. Pauli, Zur Quantenmechanik des magnetischen Elektrons. Z. Phys. **42**, 601–623 (1927)
114. R. Penrose, *The Road to Reality. A Complete Guide to the Laws of the Universe* (Alfred A. Knopf, New York, 2005), xxviii+1099 pp.
115. C. Perwass, *Geometric Algebra with Applications in Engineering*. Geometry and Computing, vol. 4 (Springer, Berlin, 2009)
116. H.-J. Petsche, *Grassmann*. Vita Mathematica, vol. 13 (Birkhäuser, Basel, 2006), xxii+326 pp.
117. L.S. Pontryagin, *Topological Groups*. Russian Monographs and Texts on Advanced Mathematics and Physics, 2nd edn. (Gordon and Breach, New York, 1966). Translated from the Russian by Arlen Brown. xv+543 pp.
118. I.R. Porteous, *Topological Geometry*, 2nd edn. (Cambridge University Press, Cambridge, 1981) (1st edn., 1969)
119. I.R. Porteous, *Clifford Algebras and the Classical Groups* (Cambridge University Press, Cambridge, 1995)
120. M.M. Postnikov, *Leçons de géométrie: Groupes et algebres de Lie* (Éditions Mir, Moscou, 1985) (Translation of the 1982 Russian edition, by D. Embarek)

121. C. Procesi, *Lie Groups. An Approach Through Invariants and Representations*. Universitext (Springer, New York, 2007)
122. M. Riesz, *Clifford Numbers and Spinors*. Fundamental Theories of Physics, vol. 54 (Kluwer Academic Publishers, Dordrecht, 1997). An edition by E.F. Bolinder and P. Lounesto of M. Riesz *Clifford numbers and spinors*. Lecture Series No. 38k Institute for Fluid Dynamics and Applied Mathematics, University of Maryland (1958)
123. P. de M. Rios, E. Straume, *Symbol Correspondence for Spin Systems* (Birkhäuser, Basel, 2014)
124. W.A. Rodrigues Jr., E.C. de Oliveira, *The Many Faces of Maxwell, Dirac and Einstein Equations*. Lecture Notes in Physics, vol. 922, 2nd edn. (Springer, Berlin, 2016)
125. D.H. Sattinger, O.L. Weaver, *Lie Groups and Algebras with Applications to Physics, Geometry and Mechanics*. Applied Mathematical Sciences, vol. 61 (Springer, New York, 1986)
126. M. Schottenloher, *A Mathematical Introduction to Conformal Field Theory*. Lecture Notes in Physics, vol. 759 (Springer, New York, 2008). A much enlarged second edition appeared in 2008
127. I. Singer, J.A. Thorpe, *Lecture Notes on Elementary Topology and Geometry*. Lecture Notes in Mathematics, vol. 388 (Springer, New York, 1967)
128. J. Snygg, *Clifford Algebra—A Computational Tool for Physicists* (Oxford University Press, New York, 1997)
129. J. Snygg, *A New Approach to Differential Geometry Using Clifford's Geometric Algebra* (Birkhäuser, Boston, 2012)
130. G. Sommer (ed.), *Geometric Computing with Clifford Algebras: Theoretical Foundations and Applications in Computer Vision and Robotics* (Springer, Berlin, 2001)
131. S. Sternberg, *Group Theory and Physics* (Cambridge University Press, Cambridge, 1994) (paperback 1995)
132. J. Stillwell, *Mathematics and Its History*. Undergraduate Texts in Mathematics (Springer, New York, 2010)
133. O.C. Stoica, The standard model algebra (2017). arxiv.org/pdf/1702.04336
134. J. Stolfi, *Oriented Projective Geometry* (Academic, New York, 1991)
135. I. Todorov, Clifford Algebras and Spinors. Bulg. J. Phys. **38**, 3–28 (2011)
136. G.F. Torres del Castillo, *3-D Spinors, Spin-Weighted Functions and Their Applications*. Progress in Mathematical Physics, vol. 32 (Birkhäuser, Boston, 2003), 246 pp.
137. G. Trayling, W.E. Baylis, A geometric basis for the standard-model group. J. Phys. A Math. Gen. **34**(15), 3309–3324 (2001)
138. L.W. Tu, *An Introduction to Smooth Manifolds*. Universitext, 2nd edn. (Springer, New York, 2011)
139. V.S. Varadarajan, *Lie Groups, Lie Algebras, and Their Representations*. Graduate Texts in Mathematics, vol. 102 (Springer, Cham, 1974)
140. V.S. Varadarajan, *Supersymmetry for Mathematicians: An Introduction*. Courant Lecture Notes, vol. 11 (American Mathematical Society, Providence, 2004)
141. J. Vaz Jr., R. da Rocha Jr., *An Introduction to Clifford Algebras and Spinors* (Oxford University Press, Oxford, 2016)
142. A. Weil, Correspondence, by R. Lipschitz. Ann. Math. **69**, 242–251 (1959). Reproduced in the second volume of A. Weil's *Œuvres Scientifiques, Collected Papers*, 556–561
143. H. Weyl, *Gruppentheorie und Quantenmechanik* (Hirzel, Leipzig, 1928). Revised edition 1931, reprinted by Dover 1950
144. H. Weyl, *The Classical Groups: Their Invariants and Representations* (Princeton University Press, Princeton, 1939). 2nd edition, with supplement, 1953
145. S. Xambó, Escondidas sendas de la geometría proyectiva a los formalismos cuánticos, in *El legado matemático de Juan Bautista Sancho Guimerá*, ed. by D. Hernández-Ruipérez, A. Campillo (Real Sociedad Matemática Española & Ediciones Universidad de Salamanca, Salamanca, 2016), pp. 233–274. https://mat-web.upc.edu/people/sebastia.xambo/GA/2015-Xambo--EscondidasSendas-JBSG-in-memoriam.pdf

146. S. Xambó-Descamps, *A Clifford View of Klein's Geometry*, 2009. Slides of the invited lecture delivered at the "International Conference on Didactics of Mathematics as a Mathematical Discipline" held 1–4 October 2009 in the University Madeira, Funchal, Madeira Island. https://mat-web.upc.edu/people/sebastia.xambo/K2/K2-Xambo.pdf
147. S. Xambó-Descamps, Álgebra geométrica y geometrías ortogonales. La Gaceta, **19**(3), 559–588 (2016). A slightly larger version in English is available in pdf here
148. S. Xambó-Descamps, From Leibniz' *characteristica geometria* to contemporary Geometric Algebra. Qüaderns d'Història de l'Enginyeria **16**(1), 109–141 (2017)
149. S. Xambó-Descamps, *17th "Lluís Santaló" Research Summer School: Geometric Algebra and Geometric Calculus with Applications in Mathematics, Physics and Engineering*, 22–26 August 2016 (Santander). https://mat-web.upc.edu/people/sebastia.xambo/santalo2016/. Invited speakers: David Hestenes (Arizona State University), Joan Lasenby (Department of Engineering, Cambridge University), Anthony Lasenby (Cavendish Laboratory, Cambridge University), and Leo Dorst (Informatics Institute, Amsterdam University)
150. S. Xambó-Descamps, J.M. Parra, Preface. Adv. Appl. Clifford Algebr. **27**, 345–349 (2017)

Index

S. Xambó-Descamps, *Real Spinorial Groups*, SpringerBriefs in Mathematics,
https://doi.org/10.1007/978-3-030-00404-0

V

W

Printed in the United States
By Bookmasters